Understanding Environmental Issues

Edited by
Susan Buckingham and Mike Turner

Los Angeles • London • New Delhi • Singapore

SAGE Publications Ltd
1 Oliver's Yard
55 City Road
London EC1Y 1SP

SAGE Publications Inc.
2455 Teller Road
Thousand Oaks, California 91320

SAGE Publications India Pvt Ltd
B 1/I 1 Mohan Cooperative Industrial Area
Mathura Road, Post Bag 7
New Delhi 110 044

SAGE Publications Asia-Pacific Pte Ltd
33 Pekin Street #02-01
Far East Square
Singapore 048763

Library of Congress Control Number: 2007936408

British Library Cataloguing in Publication data

A catalogue record for this book is available from the British Library

ISBN 978-1-7619-4235-1
ISBN 978-1-7619-4236-8 (pbk)

Typeset by C&M Digitals (P) Ltd., Chennai, India
Printed in Great Britain by The Cromwell Press Ltd, Trowbridge, Wiltshire
Printed on paper from sustainable resources

In memory of

Brian Plummer 1934–2007

Much loved colleague and teacher of geography and environmental issues, whose care for students was legendary.

Contents

Notes on Authors

Susan Buckingham is senior lecturer in Geography and Environmental Issues at Brunel University. She has written and edited a range of books on environmental issues, including 'Gender and Environment', and has designed and run modules in environmental issues and in applied geography and environmental issues at undergraduate and postgraduate level. Susan's main research investigates the links between gender and environmental inequalities. She is committed to the practical application of academic knowledge in socially just and relevant ways, and as well as lecturing and researching, she is a regular broadcaster on BBC Radio 4's 'Home Planet' and an active trustee of the Women's Environmental Network.

Phil Collins is a lecturer in Geology and Geotechnical Engineering at Brunel University, where he is currently course director for 'Civil Engineering with Sustainability'. His research focuses on environmental change, including the impact of human activities, and he has designed and run undergraduate and postgraduate modules in geomorphology, biogeography, conservation and research methods.

Adrian Combrinck is an independent GIS and Environmental Consultant, currently based in Uganda. He is also an associate of the Centre for Human Geography at Brunel University. He specialises in applied and participatory GIS and has worked in the NGO, research and government sectors in the UK and South Africa.

Katherine Donovan is a research student undertaking doctoral studies in the School of Earth, Ocean and Environmental Studies at the University of Plymouth on the theme of cultural responses to geophysical risk in Indonesia. Katherine's PhD research explores how indigenous communities around the active volcanic centres of Mt Merapi (Java) and Mt Agung (Bali) respond to emergency events and engage with disaster preparedness. More broadly, she has a strong interest in the educational outreach aspects of hazard mitigation.

Andrea Revell is an associate lecturer in the Centre for Human Geography at Brunel University currently researching and writing in Bangalore, India. Her main research areas are ecological modernisation theory, sustainable production and consumption and business and sustainability. Andrea is also involved in pedagogical research and has recently researched and published on what makes lectures unmissable.

Iain Stewart is a lecturer in geology in the School of Earth, Ocean and Environmental Studies at the University of Plymouth. His research interests are specifically in the fields of earthquake and tsunami studies, and broadly in the interdisciplinary realms of geohazards and risk science. In recent years he has become especially interested in societal impacts of abrupt geological change, having in 2003 been appointed co-leader of a 5-year geological research programme on Environmental Catastrophes and Human History. He is increasingly engaged in geosciences communication issues, mainly through his involvement with BBC science in television programmes such as 'Earth – The Power of the Planet', 'Journeys into the Ring of Fire' and 'Journeys From The Centre Of The Earth'.

Iris Turner is an associate of the Centre for Human Geography at Brunel University and has taught Geography, Environmental Pollution Science, Chemistry and Horticulture in various higher education institutions. For nearly thirty years she has taught Soil Science at the Royal Botanic Gardens, Kew and in 2004 she was awarded the Kew Medal for outstanding services to Kew. Iris's research interests have investigated the role of nitrogen and its various compounds in atmospheric, water and land pollution. She has worked extensively with mature students and has been instrumental in the development and assessment of their prior experiential learning for academic qualifications. Outside academia Iris is a very keen organic vegetable gardener and tries to practise environmental sustainability in her domestic life.

Mike Turner is an associate of the Centre for Human Geography at Brunel University but has also been a school teacher and teacher educator as well as an academic geographer. He has retained his interest in geographical education through close links with the Institute of Education in the University of London, including membership of various research project teams on the Institute's behalf. His research interests have mainly been in environmental aspects of urban planning, particularly in Mexico City but he has long been concerned with the methodology of teaching technological aspects of geography (including GIS and remote sensing) and has produced distance learning modules in these areas.

John Woodward is now a senior lecturer in the School of Applied Sciences at Northumbria University, following research and teaching posts at the University of Leeds, Brunel University and the British Antarctic Survey. His research interests are focused on ice/sediment/water interactions at the bed of polar glaciers and ice sheets to assess ice sheet stability. This involves the application of a variety of geophysical techniques including ground-penetrating radar, ground-based and airborne radar and reflection seismics. John lectures on modules in physical geography, including glaciology and climate change.

Robert Wright is a member of the faculty at the Hawaii Institute of Geophysics and Planetology. He has published many papers on remote sensing, specialising in infrared radiometry, and serves on several NASA science teams. Robert teaches a graduate/undergraduate level course in remote sensing at the University of Hawaii.

Preface and Acknowledgements

The genesis of this book was in modules on environmental issues which the Brunel University geography team taught from 1997 and which the collective authorship broadly reflects. All the contributing authors (with the exception of Katherine Donovan) have taught at Brunel during the past 10 years, and this continuing collaboration is testament to the great atmosphere of collegiality in the department. This encouraged human and physical geographers to work closely together, and through this we have come to understand each other's perspectives on a range of environmental issues, and to learn a great deal more about the issues themselves. The notes on authors included in the book illustrate the range of activities these colleagues have gone on to do – from presenting the hugely successful 'Earth – The Power of the Planet', 'Journeys into the Ring of Fire' and 'Journeys from the centre of the Earth' series on the BBC, which has brought Iain Stewart – and geography – well deserved popular recognition, to working for the British Antarctic Survey (John Woodward), and the Hawaii Institute of Geophysics and Planetology at the University of Hawaii (Robert Wright). The modules which were the starting point for this book were characterised as much by *how* they were taught as by *what* was taught in them. While this characterisation is hard to capture in the writing about environmental issues, we hope we have communicated the importance of both applied, practical learning, and 'triangulating' knowledge and recognising the perspectives from which people communicate. Much of the learning took place in groups, and some of these groups took on 'live' problem solving with NGOs, as some of the examples in the book illustrate. Many guest lecturers from a wide field have made presentations to students: from leading environmental sceptics, to environmental campaigners, politicians, external technical specialists and the Chair of a Government Commission on safe radioactive waste disposal. This mix of the intellectual and practical, group learning and keynote presentations from leaders in the field has inspired many students to move into jobs and graduate research in the environmental field.

As the two lead authors on the book, we would like to thank these colleagues for both their own contributions and the valuable feedback they have given us on our own work. We would also like to thank Robert Rojek at Sage for identifying the potential for translating these modules into a textbook in the first place, and for his enthusiasm, moral support and forbearance when the manuscript was delayed. Thanks also for the superb technical help from Sarah-Jayne Boyd and Vanessa Harwood.

But most of all, we would like to thank 10 cohorts of students who have been a privilege to work with, and from whom we have learnt probably as much as they have from us. This is the next generation who are really going to have to grapple with difficult and complex problems, and who will need sharp political understanding and a sense of social and environmental justice if these problems are going to be resolved effectively and equitably. We would like to dedicate this book to them.

1 Introduction

Susan Buckingham and Mike Turner

The ultimate purpose of this book goes beyond the learning outcomes that are to be found at the beginning of each chapter. Those outcomes are, of course, important, but at a deeper and more enduring level its purpose is to stimulate. To stimulate, that is, a lasting interest in environmental issues; to stimulate critical thinking, through which to develop your own views on the issues; and, hopefully, to stimulate a resolve to do something about it. After all, it is *your generation* that will probably decide the future of this planet.

Environmental issues are produced by the interaction of societies and the environments they inhabit, a relationship which will be explored at some length in Chapter 2. To really grasp these issues, it is necessary to understand both the physical dimensions of the 'natural' environment, as well as the social (including the economic and political). To this end, the case studies are each written by a partnership of a social scientist and a physical scientist.

During the final years of preparing the book, there has been unprecedented media coverage of environmental issues. While overseas problems aren't prominently reported in the UK, and particularly in the US, media a procession of events of unusual magnitude have caused newspapers to publicly question the role of climate change in causing, for example, Hurricane Katrina which devastated New Orleans in 2005 and the extreme flooding of southern England in the summer of 2007. The environment has become headline news. For example, *The Sun* – one of the UK's popular tabloids – had stories on its website on 8 August 2007 headlined: 'Go Green with the Sun'; 'Patio heaters poisoning planet' and 'Do your bit for the planet. Double Whammy! Help save the planet and revive your sex life with our guide' (*The Sun*, 2007). Growing concern about these issues prompted former Prime Minister Tony Blair to declare climate change 'as a greater threat to the world than terrorism' in 2005 (Adam, 2005), and has caused the UK government to recommend climate change as required learning in schools. Sustainable development is already part of the national education curriculum, more of which follows.

This book concentrates on the processes through which environmental issues are produced, understood, and consequently dealt with, and the following three chapters consider in some detail the ways in which different philosophies, value systems, or economic and political systems frame our understandings and action.

The subsequent case study chapters on food, waste, climate change and hazards are designed to illustrate how these social systems, or structures, interact with the natural environment and the implications of this. The chapter on geoinformation technology and the environment shows how recent powerful information and communications technologies enable us to know so much more about the environment, and how this changes how we understand and treat it. Finally, a concluding chapter on Mexico City draws processes and issues together in a particular geographical context.

How societies define environmental issues in the first place is critical to how decisions are made on how to deal with them, and this is contingent on past and present ideologies, or ways of thinking about the world, and there are many, although some are much more powerful than others. These ideologies come from religious beliefs and practices, science, philosophy and politics and govern not only societies' relationship with nature (which, as Chapter 2 will show, in the West is perceived as external to society, but which in many surviving indigenous people's culture is perceived as integral to it), but also our relationships with each other: between the poor and the rich, women and men, colonised and colonising.

Discourse

One of the key themes underpinning this book is that a society's relationship with its environment is a product of how powerful and influential groups in that society create, control and maintain knowledge. The ways in which these bodies of knowledge are naturalised (that is, become accepted as normal and generally uncontested by the majority of people) are myriad. In the UK, for example, what is taught in state schools is prescribed by a national curriculum, which determines the key bodies of knowledge which all children are expected to have acquired by a particular age. As this book was being written, sustainable development was included as a mandatory cross-curricular theme, while the UK government had stated that it expected all school children to learn about climate change. It is worth reflecting on how sustainable development and climate change were taught in the school you attended, and what assumptions were made in presenting these. Other mechanisms by which knowledge is naturalised into a discourse include mass media and, increasingly, the Internet, although blogging and social networking sites may dilute the development, or progression, of discourses. This is another point to reflect on.

One of the late twentieth century's key thinkers, Michel Foucault, argued that discourse is created through the exercise of power which 'perpetually creates knowledge and, conversely, knowledge constantly induces effects of power' (Foucault, in O'Tuathail et al., 1998: 3). Dalby explains how the development of environmental problems, such as biodiversity loss, ozone depletion and climate change, as global concerns have become new discourses which structure the way in which we 'know' the world. 'People and societies are constructed as "threats" or in need of "management" [and] are not merely technical issues requiring research, analysis and coordination by appropriately qualified experts' (Dalby, 1998: 180).

In the reader on geopolitics which Dalby's chapter introduces, he distinguishes between the actual physical materiality of environmental degradation and the ways in which 'they are described and who is designated as either the source of the problem, or the provider as the potential solution to the problem' (Dalby, 1998: 180). Following Foucault, Escobar has shown the importance of interrogating discourse (or the ways in which social realities are represented) to reveal how 'lived reality' is inseparable from the ways in which that 'reality' is portrayed. He goes on to explain the centrality of language to discourse, in which it is an active agent in constructing reality, and 'the process through which social reality inevitably comes into being' (Bryant, 2001: 162).

As both Chapters 2 and 3 will show, sustainable development and environmentalism have become dominant discourses in the West, in large part because they frame environmental problems and solutions in ways which neo-liberal capitalism can manage to its own benefit. So while former Prime Minister Blair has claimed that climate change is the greatest threat facing humanity, there appears to be a lack of imagination in how humanity deals with it. Rather than a reassessment of the fundamental ways in which we organise our social, political and economic lives, the emphasis is on proposed technological solutions to these, including biofuels and nuclear power, which, as later chapters show, have their own considerable negative environmental impacts.

Sustainable development

Sustainable development has emerged as a term which is used increasingly freely, although in reality it is a deeply problematic term which different interest groups use in different ways to serve their own purposes. After all, its very coinage was a compromise between the development imperatives of business, and of countries in the global south heavily reliant on their natural resources for foreign exchange, with environmental conservation interests in the West. Shiv Visvanathan illustrates how 'sustainable' and 'development' are two concepts which 'belong to different, almost incommensurable, worlds...Sustainability is about care and concern; it speaks the ethics of self restraint. It exudes the warmth of locality, of Earth as home. Development is a genocidal act of control. It represents a contract between two major agents, between the modern nation state and modern Western science' (Visvanathan, 1991). We are not saying that Visvanathan is necessarily right, but that the undeniable ambiguity of the term 'sustainable development' is not much in evidence in its everyday coinage. Such is how discourses are constructed and maintained.

That sustainable development has become a common language among businesses and governments suggests that it has the potential to offer technological fixes sufficient to maintain business as usual. It can be no coincidence that climate change has achieved such prominent attention from politicians and technologically advanced businesses, just at the time that alternatives to fossil fuels are becoming feasible, and that there is significant money to be made in green retailing. *The Independent on Sunday* reported an interview with Richard Branson in

September 2006, in which he declared himself a convert to believing in the impact of anthropogenic climate change and announced a ten-year, $3 billion investment in alternatives to fossil fuels, particularly biofuels, and has invested substantially in seven biofuel refineries in the west of the USA (Lean, 2006). Such investment clearly allows the airline industry, a major target of environmental campaigning groups, to continue to expand. This echoes the success of the phasing out of chlorofluorocarbons (CFCs) achieved in the 1980s after their use was scientifically linked to the erosion of the atmospheric ozone layer. Indeed, Jordan (1998) has shown how the enthusiasm (or lack of it) which various countries, such as France, Germany, the UK and the USA, showed towards supporting international legislation, banning first CFCs and then HCFCs, directly mirrored the development of alternative chemical substances developed by chemical companies in the respective states.

Such illustrations show the power of business and industry in shaping action on environmental issues, at least in part by subscribing to the use of an environmentally friendly sounding name – sustainable development, even if the profile of these issues is raised, initially by scientists, and more generally by environmental and health campaigning groups.

Environmental justice

This raises issues about who bears the costs of environmental damage, and another theme running through the book concerns the inequity of this at various scales, from local neighbourhoods to international divisions between the Global North and South. With the increasing economic power of countries like China and India, and to a lesser extent, Mexico and Brazil, which expands their own environmental footprints (see Box 1.1), it becomes less meaningful to divide the world in this way. However, the economic and technological power of advanced industrial nations of the Global North, and their political strength consolidated in organisations such as the G8[1] and World Bank (in which voting power is relative to size of funding), means that it is still relevant to do so.

In Chapters 2 and 3, some time is spent on explaining environmental justice as both a critique of existing social processes, and as a political movement, or campaign. This dimension of environmental issues is then taken up by the case study chapters to show how people and communities who are poorer, or otherwise discriminated against because of their ethnicity, age or gender, generally suffer environmental problems disproportionately. Indeed, 'routine' injustices which have an impact on health, such as poor housing, poor diet and food availability, living close to noxious activities such as waste disposal sites or polluting factories, coupled with a lack of control over decisions which affect their lives, weakens their ability to deal with catastrophic disaster.

Both environmental justice movements and strategies for alternative livelihoods (particularly those advocated by 'deep green' philosophies) provide some strategies for claiming a better life, and the book's authors hope that reading through the chapters, and thinking through some of the questions they pose, will enable readers to think imaginatively about some of the solutions to what are intricately linked social/environmental problems.

Box 1.1 Ecological footprinting

Mathis Wackernagel, one of the original developers of this concept, stresses the strength of ecological footprinting as enabling planners, individuals and communities to visualise human impact on the Earth in order to begin to minimise this. He admits that it is not a precision tool, but one which offers a way of measuring ecological sustainability by calculating the resources people consume, the waste they generate and the biologically productive area needed to provide enough space for this (Wackernagel, 1998). In an assessment of ecological footprinting, Levett has argued that scoping the contributing factors of the footprint are essentially political and value-laden activities and warns against it becoming a mechanistic planning tool, however, he recognises the concept's usefulness as a framing device (Levett, 1998). There are now many different schemes in which people can calculate their personal ecological footprint, some of the more reputable including those used by the World Wildlife Fund (WWF) and the Royal Society for the Encouragement of Arts, Manufactures & Commerce (RSA). The Stockholm Environment Institute claims that it has created the 'first consistent method to calculate comparable Ecological Footprints for every Local Authority in the UK' (Stockholm Environment Institute, 2007). It has calculated the world average ecological footprint as 2.2 hectares per person: an 'ecological overshoot' of 21 per cent, given that there is only 1.8 hectares per person of biologically productive land available. Communities in the Global North are particularly expansive users of this land, with a study in Nottingham, England, estimating that each person's ecological footprint is 5.3 hectares, marginally less than the UK average reported by the WWF in 2006, but a clearly unsustainable size (Birch et al., 2005). The WWF in its *Living Planet Report* shows the ecological footprint of the USA to be 9.6 hectares per person; by comparison, China's is reported as 1.6 (WWF International, 2006).

Student involvement in environmental issues

The genesis of this book was a series of modules on environmental issues in undergraduate and postgraduate taught courses in Geography, Earth Sciences and Environmental Change. The research and teaching expertise of the lecturers (who are the contributing authors) has largely determined the choice of case studies. This choice inevitably excludes some important issues, such as energy and water, but those which are included – food, waste, climate change and hazards – are all important, of increasingly high profile, and need to be understood in their complexity if the problems which the authors have identified are to be addressed. Moreover, understanding how these issues are produced and dealt with will enable readers to develop their skills in understanding other environmental problems and their impacts.

As well as a commitment to the themes identified in this introduction, our teaching of environmental issues has also been influenced by a commitment to engaging students in researching real-life issues for themselves. This has been achieved in partnership with a number of organisations which Box 1.2 illustrates. Such engagement requires students (and their tutors) to understand a range of assumptions made about environmental issues, and the different perspectives adopted by the environmental, non-governmental organisations (ENGOs) worked with. While this book will involve students in a different, less physical, kind of engagement, we hope that it will stimulate an intellectual engagement to reflect on their own assumptions and perspectives, those of their tutors and lecturers and of politicians, broadcasters and other opinion-formers. Not least, we hope that it will stimulate readers to take action that will, itself, have a positive impact on their present and future environments, in the broadest sense.

Box 1.2   Examples of student community environmental partnerships

Example 1: As a second year undergraduate, Katie worked with the Groundwork Trust (GT) to develop an accessibility guide for visually and mobility impaired users of a large stretch of managed countryside in West London. With a fellow student, Katie surveyed and evaluated the area, producing a pamphlet which is available from the Visitors' Centre. As a result of working with the GT, Katie applied her Level 3 major research project to developing a food growing project for refugees on local unused allotments. Working with the GT, the local authority and a number of refugee organisations, Katie produced a proposal for the GT and the authority to develop.

Example 2: Four second level undergraduate students worked with a London borough on developing a system for distributing otherwise wasted energy to council properties, including social housing, through a local grid. The idea was inspired by the pioneering borough of Woking in southern England which, as Chapter 7 shows, has achieved significant carbon savings and reduced energy costs for households on low income. The group of human geographers worked with a team of student engineers from an American university, which incorporates three months of applied community practice as part of their degree programme (including projects such as the evaluation of tsunami-proof housing in Thailand, and AIDS prevention education in Namibia).

Example 3: As part of the assessed curriculum on the Applied Environmental Research module, a postgraduate student worked with the Women's Environmental Network (WEN) researching climate change and its likely impact on women. This formed the basis of the Climate Change Manifesto for Women, launched by WEN and the National Federation of Women's Institutes in 2007. Another student on the same course designed, conducted and analysed a survey of women's views on nuclear power for WEN's response to the Government's consultation on future energy planning in the UK.

Note

1. The G8 comprises: Canada, France, Germany, Italy, Japan, Russia, the UK and the USA.

References

Adam, D. (2005) 'Environmentalists tell PM: don't abandon global warming fight', *The Guardian*, 2 November 2005.

Birch, R., Wiedmann, T. and Barrett, J. (2005) *The Ecological Footprint of Nottingham and Greater Nottinghamshire*. York: Stockholm Environment Institute.

Bryant, R. (2001) 'Political ecology: a critical agenda for change?', in N. Castree and B. Braun (eds), *'Social Nature: Theory, Practice and Politics'*. Oxford: Blackwell.

Dalby, S. (1998) 'Introduction to environmental geopolitics', in G. O'Tuathail, S. Dalby and P. Routledge (eds), *The Geopolitics Reader*. London: Routledge.

Jordan, A. (1998) 'The ozone endgame: the implementation of the Montreal Protocol in the United Kingdom', *Environmental Politics*, 7(4): 23–52.

Lean, G. (2006) 'The Jolly Green Giant', *The Independent on Sunday*, 24 September 2006.

Levett, R. (1998) 'Footprinting: a great step forward, but tread carefully – a response to Mathis Wackernagel', *Local Environment*, 3(1): 67–74.

O'Tuathail, G., Dalby S. and Routledge, P. (eds) (1998) *The Geopolitics Reader*. London: Routledge.

Stockholm Environment Institute (2007) http://www.regionalsustainability.org (accessed 9 August 2007).

The Sun (2007) http://www.thesun.co.uk/sol/homepage/news/special_events/green_ week/

Visvanathan, S. (1991) 'Mrs Brundtland's disenchanted cosmos', *Alternatives* 16(2).

Wackernagel, M. (1998) 'The ecological footprint of Santiago de Chile', *Local Environment*, 3(1): 7–26.

WWF International (2006) *Living Planet Report*. Gland, Switzerland: WWF International.

SECTION 1
FRAMEWORKS

2 Approaching Environmental Issues

Susan Buckingham

Learning outcomes

Knowledge and understanding:

○ you will understand that environmental problems have their origin in ideological structures and are the result of particular decisions made by, generally, societies' most powerful interests

○ you will recognise sustainable development as a contemporary discourse which is inherently problematic, be able to distinguish between different approaches to environmental problems and their origins, and develop an understanding of selected critiques of prevailing environmental discourses

Critical awareness:

○ through reflecting on your own environmental views and behaviour, you will recognise that strategies for dealing with environmental problems are dependent on how these problems themselves are understood and produced

---

Introduction

The environment at any point in time is a product of the interplay between social, physical and natural/biological processes. Given that our actions as human beings, organisations and communities are influenced by a set of cultural practices (comprising, for example, linguistic, religious, spiritual and philosophical underpinnings which inform our 'world views'), it should be clear that the environment will therefore be affected by these practices. This chapter will focus on some of these cultural foundations to demonstrate the importance of understanding the role of the values they represent in producing particular environmental problems, and the strategies for dealing with these problems.

Although societies tend to be characterised by a single set of values ('Christian capitalist' America, the 'Islamic' Middle East or 'Communist' Cuba), in reality each dominant set of values represents a particular moment in which these have achieved 'hegemonic power' over other, possibly competing, values. While these values are mutable and fluid, they are at the same time intimately tied to dominant groups in society in whose interest it is to 'naturalise' these values, such that society thinks of them as inevitable. One term that this chapter, and book, will consistently return to is 'discourse': a practice through which a set of ideas achieves and maintains momentum, and which has already been discussed in some detail in Chapter 1 (see also Box 2.1).

Box 2.1   Defining key terms

Nature: While nature is often seen as separate from humans and society (as a resource, for example, or a physical entity which suffers from human activity), this is only one way of conceptualising it. Noel Castree (2001) argues that there are several ways of defining nature and that this conventional Western view is of an external nature, or **environment**. He reviews the ways in which critical geographers see nature as a social concept – 'social nature', in that the only way in which we can 'know' nature is through our perception of it, consequently it is internalised by our perception. Also, because we are 'physical' bodies, we physically interact with nature, forming what Erik Swyngedouw (1999) calls 'socionature'. Through this interaction, nature is constantly 'remade' or 'reconstituted', which again suggests that it is unrealistic to think of nature as something separate from ourselves. (Castree, 2001; Swyngedouw, 1999)

Discourse: 'Sets of capabilities people have, as sets of socio-cultural resources used by people in the construction of meaning about their world and their activities. It is NOT simply speech or written statements but the rules by which verbal speech and written statements are made meaningful.' (O'Tuathail and Agnew, 1998: 80)

Hegemonic power: A position held by a state or a class when it so dominates its sphere of operation that other states or classes are forced to comply with its wishes. (Flint and Taylor, 2007)

In the physical sciences, Thomas Kuhn (1970) conceived the idea of paradigms – a set of dominant or, what could now be termed hegemonic, ideas which frame scientific understanding at any one time. Kuhn used the term 'paradigm shift' to explain how a set of theories, or dominant, or hegemonic, set of ideas, gives way to another over time. It was through paradigm shifts, Kuhn argued, that scientific knowledge develops. A theoretical body of knowledge has power only as long as it provides a persuasive explanation as to why things are as they are, but, as it is in the nature of science to question and refute, the dominant paradigm within which the theory has developed is constantly challenged. A robust theory is better able to withstand challenges but ultimately will yield to alternative theories if they are able to offer more persuasive explanations in the light of more sophisticated or extensive data. Challenges to existing theories may eventually lead to the emergence of a completely different paradigm, which will, in turn, become the dominant, or hegemonic, paradigm and itself open to challenge.

It is through challenge, then, according to Kuhn, that science develops. Within each paradigm, science is expected to proceed in a linear fashion in its attempt to uncover a pre-existing body of knowledge, although the paradigm shifts can represent radically different changes in direction of scientific thought, as the title of Kuhn's book, *The Structure of Scientific Revolutions*, suggests. Kuhn was one of the positivists working in Europe in the mid-twentieth century for whom the search for knowledge represented truths to be discovered, rather than a socially contingent way of understanding. This particular – positivist – way of considering the world sits firmly in the 'modernist' world view, which emerged out of a transformation of society, referred to as 'the enlightenment,' which Karen Armstrong argues was 'the last of the great revolutions of human experience' (Armstrong, 2005: 119).

The enlightenment, the Christian tradition and the environment

The enlightenment – a term which, itself, reveals the expectations of its protagonists – has had profound implications for nature and the environment, as it was founded on 'an economy that seemed, potentially, to be indefinitely renewable' (Armstrong, 2005: 120). Philosophical approaches which characterised the enlightenment were powerful shapers of industrial society. Attitudes which have defined features of Western industrial, and post-industrial, societies – for example, nature's 'use' value as a present or potential resource, and allocations based on rights (even though the groups to whom rights have been extended vary over time, as the later section on environmental justice will show) – can clearly be traced back to the enlightenment.

According to Eric Hobsbawm, the enlightenment was defined by a conviction of the progress of human knowledge, rationality, wealth, civilization and control over nature (Hobsbawm, 1977: 54). The enlightenment drew some of its inspiration from Greek philosophy, for example, Aristotle's focus on husbandry and nature as a resource for the careful use of society. Because Christianity, too, was inflected with Hellenism (Chrysaviggis suggests that the early Christian tradition was influenced by Greek interest in the cosmos, 2006: 94), it is not surprising to find echoes of the Christian articulation of society with nature in enlightenment thinking. Four themes can usefully be identified in Christianity, which also find their way into enlightenment thinking. These should, however, be seen as interwoven: dualism, by which a division between humans and nature is established; hierarchy, by which humans are placed at the apex of living beings; utility, through which nature is seen as a resource for human beings; and stewardship, by which humans are, through their superior intelligence, charged with taking care of nature for both lesser beings and future generations, prefiguring the central precept of 'sustainable development'.

Hobsbawm insists, however, that the enlightenment represents a departure from religious traditions which, 'for most of history and over most of the world (China perhaps being the main exception) the terms in which all but a handful of educated and emancipated men thought about the world were those of traditional religion' (Hobsbawm, 1977: 266). By the late eighteenth century, the enlightenment had become the 'religion' itself. In his book, *The Age of Revolution*, Hobsbawm suggests that enlightenment thinking had created a secular, rationalist and progressive individualism which sought to free society from the 'ignorant tradition of the Middle Ages and superstition of the churches' (as distinct from 'natural' or 'rational' religion). This thinking formed the basis of the French and American revolutions founded on an emerging bourgeois and capitalist middle class.

Armstrong's (2005) particular analysis of the enlightenment concerns the place of myth in society, which she describes as 'stories that enabled us to place our lives in a larger setting, that revealed an underlying pattern and gave us a sense that . . . life had meaning and value' (p. 2). Mythology, as a strategy for helping people cope with the 'problematic human predicament' is, she argues, both psychology and religion and is continually adapted to suit changing conditions (although it also helps to structure those conditions).

Nature is intrinsically part of the myths societies sustain, and the role of nature relative to human beings and other creatures in those myths will influence the way in which societies view and treat it. For example, as this chapter will show, Christian societies, which see human beings as stewards of nature, on behalf of less elevated forms of life and future generations, consequently view nature quite differently from Buddhists, for whom nature is an interconnected web of which humans are an integral part. These traditions, and the contemporary influence they have on both economic and environmental thinking, will be examined in more detail later in this chapter. It is not therefore surprising that Western society, philosophically grounded in Christian and scientific world views, combines a view of nature which it both 'stewards' and controls, in various measures at various times. Arguably, sustainable development represents an extension of this combined world view, as it seeks to articulate environmental considerations with sustained economic development, a position many critics suggest is untenable but which, this book argues, appeals to enough powerful interests to have elevated it to the status of a prevailing discourse.

Returning to Kuhn's paradigms, these are useful in showing that the dominant way of understanding nature or the environment is partly a result of an ability to answer critics by the power and salience of an argument which accords with the prevailing norms of the society in which it is developed. However, discourse theory shows how the power of any particular way of looking at the world derives also from the power of its proponents to shape the way in which individuals learn, read and think about the world. It is instructive to consider the ways in which alternative environmental understandings are treated to see how they do – or do not – influence the prevailing world view.

Consider, for example, North American or Australian societies in which the prevailing Christian-scientific world view has overwhelmed the values of indigenous communities, which remain only in pockets of these societies. As well as existing in their own right, these indigenous understandings of the relationship between humans, non-human beings and nature have inspired a number of radical ecocentric environmental movements. Both indigenous and radical 'alternatives' are tolerated by Western society, which prides itself on this tolerance and its democracy, but are constructed as minority interests with limited practical capability and, as such, can be effectively marginalized. This is particularly true of the current debate on climate change which, in the West, is focusing on technological strategies for amelioration such as the development of biofuels, hydrogen power, and greater energy efficiency to reduce carbon dioxide pollution, rather than engage with alternative approaches to the ways in which we live, work, travel and consume. However, this is inherently problematic.

Biofuels (derived from plants such as oilseed rape, sugar beet and wheat) are currently being heralded as the 'carbon-neutral' alternatives to fossil fuels, as they consume enough carbon dioxide in growing to offset that released when burnt as fuel. However, this can be misleading. Because of the nitrogen fertilisers used in producing the oilseed rape from which some biofuels are derived, significant amounts of carbon dioxide (7 kilograms of CO_2 for every kilogram of nitrogen fertiliser made) are produced before the crop is even harvested (Maynard, 2007: 30); 'they also create more carbon dioxide by transportation than they save by reducing fossil fuels' (Young, 2006). In addition to creating CO_2 emissions, biofuel production causes a number of other problems, which tend to impact most heavily on the poor, including soil erosion, loss of land for food growing (an area larger than the entire agricultural land currently in use in the UK would be needed to replace the annual UK consumption of petroleum). The West is also outsourcing production of biofuels to countries in the Global South, resulting in the clearance of rainforest in countries like Malaysia, in which 87 per cent of deforestation between 1985 and 2000 was caused by biofuel production (Maynard, 2007: 28). As this suggests, even in the developing economies of India and China, in the past variously influenced by very different world views such as Hinduism, Buddhism, Confucianism and Communism, a Western-scientific approach now dominates development and those economies' relationship with nature, as the example of the Narmada dam project in India presented in Chapter 3 makes clear.

Business as usual, or time for a paradigm shift?

At its most extreme, this Western-scientific world view, of which sustainable development is now effectively a part, can be deemed 'cornucopian', in which progressively sophisticated technology and ingenuity can provide perceived solutions to environmental problems, whether these be resource depletion (by identifying another resource), pollution or waste (by creating mitigation technologies). In 1980, Julian Simon, one of the better known cornucopians, threw a challenge to Paul Ehrlich (a critic of Western environmental values and prominent author of, among other books, *The Population Bomb*), which invited him to name a basketful of commodities which, if they were becoming scarcer, could be expected to increase in price over an extended period of time. Ehrlich bet $1,000 on five metals – chrome, copper, nickel, tin and tungsten – in quantities that each cost $200. A futures contract was drawn up, obligating Simon to sell Ehrlich and his colleagues these same quantities of the metals 10 years later, but at 1980 prices. If the price of the basket had risen, Ehrlich would have won the argument – and the bet; alternatively, if the commodities had become cheaper, Simon would have triumphed for the cornucopians.

Q: Who do you think won the argument and why? (see Appendix 2.1)

Professor Simon was involved in another clash of views when he debated the environmental proposal 'Scarcity or abundance?' with Professor Norman Myers, in 1994. The dean of the School of International and Public Affairs at Columbia University, Professor John Gerard Ruggie, who hosted the debate, suggested that the proponents held different world views which caused 'our two authors' to assess the real world in completely different ways. Both drew on published data on life expectancy, soil degradation and biodiversity loss to 'prove' their case (in Simon's case that 'more people almost surely imply more available resources and a higher income for everyone', p. 124 and, in Myers, that 'we are at a watershed in human history because of the grand-scale environmental degradation that is overtaking our planet in conjunction with excessive population growth [is causing] mass extinction', p. 125) (Myers and Simon, 1994).

As Chapter 4 on economic values explores in more depth, the inflection of the Western world view with some environmental considerations has produced an 'ecological modernisation' approach to business, which advocates producing goods and services more efficiently, thereby saving businesses money as well as answering environmental critics. Businesses, for example, claim environmental advantages from a range of practices such as reducing packaging, substituting biodegradable materials for oil-based plastics or switching to so-called renewable fuels, meanwhile continuing to expand production and stimulate consumption. While making sound financial sense, both from expenditure savings, and in terms of enhanced publicity, this does little to assuage environmental – or social – problems in the longer term. As Hawken et al. (1999) argue in *Natural Capitalism*, as long as the economy is organised around and privileges financial, rather than natural and social resources, it will continue to be inefficient on all three counts. Moreover, continued inefficiencies in the West are used as a smokescreen to argue against development in the Global South (as the chapter on climate change makes clear when discussing how the US refusal to sign the Kyoto Protocol hinges partly on its insistence that developing economies such as Brazil, China and India should be subject to similar greenhouse gas emissions targets as Western countries). However, for all its limitations, the concept of sustainable development is valuable in that it has brought together the *three* pillars of the economic, the environmental and the social. The danger within current applications of sustainable development is that the economic has primacy and that environmental improvements are introduced when they can be seen to deliver economic benefits, or when there is sufficient fear of imminent resource depletion (see Chapters 4 and 8 on economics and climate change). The current enthusiasm for energy and industrial efficiencies is heavily predicated on the availability of replacement technologies (even though, in the case of biofuels, for example, the social and environmental costs of growing sufficient of these to maintain the world's current, let alone projected, car mobility, are potentially huge, as the chapter has already indicated), and on the economic gains that can be made from implementing these. It is telling that ex-US President Bill Clinton's endorsement of the book *Natural Capitalism* says that it 'proves beyond any argument that there are presently available technologies, and those just on the horizon, which will permit us to *get richer* by cleaning, not by spoiling, the environment' (1999: back cover; emphasis added.)

At the heart of this discussion is the question, which the reader of this book might like to reflect on, as to whether the environmental problems which have emerged as a result of the dominant Western world view can adequately be addressed within that world view, or whether a paradigm shift is required. Part of this question hinges on an analysis of exploitation: as well as natural resources being exploited by industrial society, this society has a long history of human exploitation. A number of current critiques of the Western social-environment relationship call into question the way in which disadvantaged people and communities disproportionately experience environmental problems, so that elites can continue their resource-intensive lives relatively unaffected by environmental damage. This chapter will consider in some depth two of these critiques: environmental justice and ecofeminism. Environmental justice, and some strands of the ecofeminist movement have, themselves, emerged from enlightenment discourses around rights and justice and engage these to try to redress the imbalance of environmental goods and bads. Ecofeminism goes further in suggesting that it is the exploitative nature of Western capitalism, which treats women and the environment as relatively powerless interests, that equips women particularly well to address environmental problems, as they share the experience of exploitation. Finally, while Western society is potentially modified by critiques such as environmental justice and ecofeminism, the chapter will conclude by considering the ways in which a completely different approach suggests an alternative way of living with nature. The choice of Buddhism is deliberate in that as well as providing the foundation for a way of living in some non-Western societies, it has also inspired a sizeable part of the deep green movement in the West and, as such, can be seen as having the potential to influence society–environment relations globally.

The relationships between approaches to the society–environment nexus

Broadly speaking, analyses of environmental philosophies have tended to conceptualise these philosophies on a grid of ecocentrism and anthropocentrism (or people-centredness), and between left and right politics. Extreme points on this grid (see Figure 2.1) would be eco-Marxism (left, anthropocentric), eco-fascism (right, ecocentric), deep ecology (ecocentric) and cornucopian (anthropocentric).

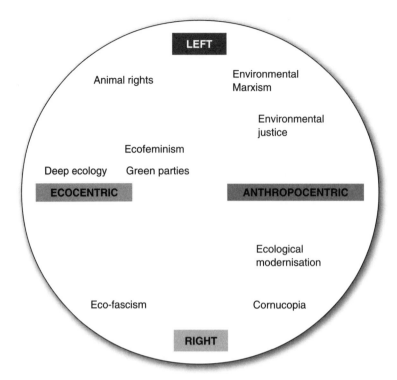

Figure 2.1   Approaches to the environment

David Pepper reviews some of these perspectives, grouping ideas under 'radical environmentalism' (in which he includes biocentrism and ecofeminism), 'modern ideas about nature and science' (covering the scientific revolution and Christianity) and ecocentrism (including utopian socialism and ecological economics) (Pepper, 1996).

As Anthony Giddens in *Beyond Left and Right* has argued, the environment is increasingly an example of a contemporary issue which breaks typical political pigeon-holing, with politicians from most political parties racing to the middle ground of ecological modernisation (Giddens, 1994). What little concern there was about environmental problems on the left was articulated as ecosocialism, drawing on the work of neo-Marxists and earlier social thinkers such as John Stuart Mill, William Morris and Henry David Thoreau (as Chapter 3 develops). In many respects, this agenda has been subsumed in environmental justice movements which take up the case of indigenous groups, the poor, people of colour, and people in the Global South. Andrew Dobson broadly conceptualises this amalgam of approaches as 'environmentalism', which he defines as 'a managerial approach to environmental problems, secure in the belief that they can be solved without fundamental changes in present values or patterns of production and consumption' (Dobson, 2000: 2). In these times of increasing attention to the environment, for whatever reason, it becomes increasingly difficult to differentiate distinct environmental philosophies by a left/right division, as environmental thinking permeates more widely. Green parties, while constituting formal political parties, particularly in Europe, where Green MEPs represented just under 6 per cent of Members of the European Parliament in 2004 (see Chapter 3) would argue for a more radical change in the relationship between society and nature. Two leading figures in the British Green Party, Michael Woodin and Caroline Lucas (Woodin was the Principal Speaker of the party in England and Wales until his death in 2005, while Lucas is an MEP representing south-east England) capture the social-environmental aims of the green movement. These include the reconstruction of 'patterns of human activities and relationships so that they come to respect the natural systems on which they depend', a prerequisite of which is that 'equity and social justice are woven into the fabric of society' (Woodin and Lucas, 2004: xix).

There are also anarchist politics, which, opposing the state political system, also fall outside the left/right divide. These have a long tradition which St Augustine's advocacy in the 4th and 5th centuries CE suggests: 'Augustine advocated community of goods, care for creation, and even intergenerational responsibility when he stated that Christians should regard themselves as pilgrims on a journey who paused for a while at an inn' (Hart, 2006: 68). Also, Aquinas advocated that 'possessions and provision of the necessities of life for all people' meant that a 'person in need [could] go quietly in the night to take from the barn of a nobleman what was needed for survival: what they would be taking would actually be their own property, since their need in their time and place made all property come under the natural law of common property' (Hart, 2006: 69).

Such beliefs characterise the anti-globalisation movement and a number of protests described by Naomi Klein as organised through webs and networks (Klein, 2001). Dobson includes some such groups in his concept of 'ecologism', which he defines as holding 'that a sustainable and fulfilling existence presupposes radical changes in our relationship with the non-human natural world, and in our mode of social and political life' (Dobson, 2000), more of which will be discussed later in this chapter. Dobson's distinction between environmentalism and ecologism is an important one, as it also challenges the continuum of environmental thought often described as between deep and shallow, or between dark and light green, ecology. He does so on the grounds that while ecologism is an ideology in itself, environmentalism can be ascribed to any other ideology which incorporates some environmental strategies (such as environmental capitalism or environmental socialism), without challenging their underlying foundations. No society will change ideologically by switching from fossil to biofuels, from landfilling to recycling rubbish, or by enforcing more efficient pollution controls: these are environmental adjustments which permit existing ideologies to proceed with business as usual. Indeed, it can in the short term serve to mask some of the existing ideology's failings. On the other hand, ecologism, Dobson suggests, 'challenges the political, social and scientific consensus that has dominated the last two or three hundred years of political life' (Dobson, 2000: 8).

One example of an ecologist argument is social ecology, which is arguably an intellectual, as much as a political or campaigning movement. This draws on the work of, among others, Murray Bookchin – an American earlier involved in labour politics. Social ecology argues against the dualism which characterised the enlightenment – and subsequent Western – ways of thinking, as it tends to separate nature from society and culture. Bookchin claims that the social ecology approach is diverse enough to 'nurture freedom, [is] an interactivity that enhances participation and a wholeness that fosters creativity, a community that strengthens individuality [and] a growing subjectivity that yields reason.' (Bookchin, 1990: 203). He advocates the bioregion as the basic organising principle for human societies. Social ecology, then, articulates society or culture with nature in complex holistic ways that 'harmonize[s] our relationship with nature in a creative not destructive "metabolism" with nature' (ibid.: 203). More of this later, when the chapter turns to consider deep ecology.

With the increased visibility of environmental justice movements (which focus on race as much as class or poverty) and global anti-globalisation movements advocating international links through hyper-space as much as local self-sufficiency, the kind of socialist and bioregional ways of thinking of Bookchin may appear a little dated. However, when adapted to changing circumstances, such as the growth of information communications technology, they have much potential, as the chapter will conclude. Box 2.2 quotes from Gary Snyder's consideration of the nature of the bioregion, from which can be seen that it is a more coherent way of organising and managing the way in which we live as an integral part of our environment than often arbitrary political boundaries, such as the 49th parallel, which defines the boundary between Canada and the USA, or those created in Africa by the colonial powers of Europe and the USA at the Berlin Conference in 1884 (Pakenham, 1991). It is worth considering how the bioregion described here draws from biogeographical concepts.

Box 2.2   The bioregion

'The world of culture and nature, which is actual, is almost a shadow world now, and the insubstantial world of political jurisdictions and rarefied economies is what passes for reality. We live in a backwards time . . . . Regions are "interpenetrating bodies in semi-simultaneous spaces"' (Cafard, 1989). Biota, watersheds, landforms, and elevations are just a few of the facets that define a region . . . . The coastal Douglas Fir, as the definitive tree of the Pacific North West, is an example . . . . Its northern limit is around the Skeena River in British Columbia. It is found west of the crest through Washington, Oregon and northern California. The southern coastal limit of the Douglas Fir is about the same as that of salmon, which do not run south of the Big Sur River. Inland it goes down the west slope of the Sierra, as far south as the north fork of the San Joaquin River. That outline describes the boundary of a larger natural region that runs across three states and one international border.

'The presence of this tree signifies a rainfall and a temperature range and will indicate what your agriculture might be, how steep the pitch of your roof, what raincoats you'd need. You don't need such details to get by in the modern cities of Portland or Bellingham. But if you do know what is taught by plants and weather, you are in on the gossip and can truly feel more at home. The sum of a field's forces becomes very loosely the "spirit of the place". To know the spirit of the place is to realize that you are part of a part and that the whole is made of parts, each of which is whole. You start with the part you are whole in.

'Bioregionalism is the entry of place into the dialectic of history. Also, we might say that there are "classes" which have so far been overlooked – the animals, rivers, rocks and grasses – now entering history.

'These ideas provoke predictable and usually uninformed reactions. People fear the small society and the critique of the State. It is difficult to see, when one has been raised under it, that it is the State itself which is inherently greedy, destabilising, entropic, disorderly and illegitimate . . . .

'The bioregional movement is not just a rural program: it as much for the restoration of urban neighbourhood life and the greening of the cities. All of us are fluently moving in multiple realms that include irrigation districts, solid waste management jurisdictions, long-distance area code zones, and such. Planet Drum Foundation, based in the San Francisco Bay Area, works with many other local groups for the regeneration of the city as a living place, with projects like the identification and restoration of urban creeks . . .'

(Snyder, 1990: 37–8)

Having briefly reviewed a range of environmental approaches, what the remainder of this chapter intends to do is to focus on three socio-environmental critiques of the way in which the dominant world view (of global capitalism under neo-liberalism) treats the environment, to show how these critiques have emerged from particular bodies of thought and to invite the reader to think about the likely impact these challenges will have. Elsewhere in the book, we put these, and other, critiques into particular contexts, where appropriate. These critiques are clearly not exclusive, and are not even necessarily the most important, but they illustrate key points which the book is making about how ideas and consequently processes are produced, and the relationship between philosophies. These are primarily critiques of the society–nature relationship, but it is also worth reflecting on critiques of decision-making processes. Environmental justice and some interpretations of ecofeminism argue that there is insufficient attention given to the views of people most affected by environmental problems; these critiques are more thoroughly developed in the environmental democracy discourses proposed by, for example, Burgess et al. (2000); Hajer (1995) and Mason (1999), which stress the importance of deliberative or participative democracy, in which a range of 'stakeholders' (that is, people who hold a stake in the area being developed, whether that is their home, their business or professional concern) are engaged in a participation or consultation exercise which they understand. This approach is distinct from the nominal participation exercises often run by, for example, local councils, and which rely on the participation of a small number of unrepresentative interests, or on circulated questionnaires without investing in educating the broader population about the issues at stake.

Critiques

Environmental justice

The term 'environmental justice' describes a range of social movements which originated in the USA in the 1980s, although the principles underlying this – that individuals and communities ought not to be unfairly disadvantaged in their exposure to environmental problems – have a longer history as Chapter 3 on environmental politics demonstrates. It was born out of the civil rights movements of the 1960s and specifically related to the frustrations of African-American and Hispanic communities, who experienced greater degrees of environmental pollution and other disadvantage than their income alone would predict. Bob Bullard, a seminal figure in both the movement and in its analysis, dates its inception to the explosion at a hazardous waste disposal site which received PCBs, among other chemical waste, in 1982 in Warren County, South Carolina. This event, and the protest which ensued, focused campaigning groups' attention on the relationship between minority ethnic status and health and environmental problems. In 1987, the Commission for Racial Justice presented evidence that race was the most potent variable in predicting where a factory would be located which led, ultimately, but not until the Democratic Clinton administration in 1992, to a review of federal legislation. An Executive Order passed in 1994 ensured that federal actions were taken to address environmental justice in minority populations and low income populations (Bullard, 1999). Despite a series of legislative measures, however, there continue to be significant violations eloquently documented by writers such as Melissa Checker (see Chapter 7 on waste) and Hilda Kurtz. Kurtz, for example, analyses a case in Louisiana in which a multiracial poor community living in an industrial neighbourhood protested against the expansion of a chemical-producing factory. Her research catalogues the way in which the interests of government and the chemical company combined against the interests of the multiracial local community. Of particular interest to the student reader might be the role of students and their lecturers at Tulane University, in providing scientific evidence to the community to support their challenge to the siting of another chemical plant in the neighbourhood. As a result of the outcome of the case, in which the Louisiana State Government upheld the community case, the Federal Government deemed any evidence provided by students to be inadmissible in future environmental injustice cases (Kurtz, 2003).

In the UK, Friends of the Earth Scotland used environmental justice as a campaigning framework to protest against the disadvantaging of poor neighbourhoods in Scotland by the location of waste disposal sites (see Dunnion and Scandrett's 2003 analysis in Chapter 7 on waste), since when the concept has been adopted by other groups concerned about the lack of ethnic minority representation in the

UK environmental movement (such as Capacity Global and the Black Environmental Network), and left-leaning local government (the Greater London Assembly has produced a position statement on environmental justice).

At the global scale, environmental justice provides a language with which relatively powerless groups can protest against the inequities of food security (see Vandana Shiva's analysis of multinational interests in appropriating food production in Chapter 6 on food), hazardous waste (see the discussion on the Basel Convention on the shipment of hazardous waste in Chapter 7 and Chapter 10 on Mexico City), water availability (see Chapter 10) and oil exploitation (see the discussion on Shell's oil production in the Niger River Delta region of Nigeria, in Chapter 3). The groups treated unjustly here are both indigenous groups within countries in the Global South (for example, the minority Ogoni people in Nigeria experiencing pollution as a result of oil extraction in the Niger River Delta), and the Global South more generally, which has access to considerably less power than countries in the Global North, whose influence on international institutions such as the World Bank, International Monetary Fund (IMF), and World Trade Organization (WTO) is incontrovertible.

A corollary to this view follows from the increasing awareness that global biodiversity is decreasing and therefore that humankind has a responsibility for ensuring the stabilisation of the rate of decrease. This view leads to an extension of the concept of environmental justice to include global and intergenerational inequity, rather than the highly localised approach of the American tradition. Examples cited in an ESRC report on environmental justice (ESRC Global Environmental Change Programme, 2001) include the effects of the Chernobyl accident and the impacts of industrial pollution, concentrated in fish through the food chain, on the Inuit people's staple diet. The report also suggests that the effects of global warming are an injustice to people in the Global South and to future generations, neither of whom were responsible for their causation but both of whom have to suffer their consequences.

However, it is important to point out that environmental justice movements, as Schlosberg (2007) points out, do not necessarily incorporate *ecological* justice as one of their aims (where ecological justice refers to the way in which non-human nature is treated). Environmental justice has emerged from the political science tradition of 'rights' and distributive justice which, with its focus on the redistribution of environmental goods and bads, places it on the left of the political-environmental spectrum. Schlosberg argues the need for a greater attention to issues of recognition and participation alongside distribution to widen the range of environmental justice, which articulates with an increasing body of work on environmental citizenship linked to participative democracy, as this chapter has highlighted earlier, and Chapter 5 has discussed in relation to participatory GIS.

Environmental justice has a history of anthropocentrism in which the central focus is on human beings, arguing that only sentient beings can negotiate distribution, which leads to non-human nature qualifying for compassionate treatment, rather than justice in its own right. The influences on this approach can be traced back through Rawls in the 1970s to enlightenment thinking, and to ancient Greek philosophies which stressed rationality, impartiality and objectivity, as well as through the civil rights movements. Schlosberg's argument that environmental justice has neglected ecological justice (and his book *Defining Environmental Justice* seeks, as one of its aims, to bridge the gap between environmental and ecological justice) suggests that the civil rights movement, in some ways, has been used selectively. After all, Martin Luther King was a Christian minister, who was also significantly inspired by Mahatma Gandhi, whose strategy of peaceful resistance, satyagraha (from Gujarat, 'sat' for truth and 'agraha' for firmness, Gandhi, 1993: 319) was intimately derived from the Hindu (and Buddhist) philosophy of ahimsa – or non-harm. As the discussion on Buddhism will later show, some spiritual traditions outside Judaism–Christianity–Islam, would not claim that non-human nature was not sentient. The 'engaged Buddhism' of Vietnam, responding to the injustices of the colonial war in Vietnam raised global questions about environmental injustice, and demonstrates religious and alternative influences on the environmental justice movement outside the West, as do some of the liberation theology movements in Catholic Latin America.

While it has done much to raise the profile of individuals and communities in poverty, and discriminated against by colour, environmental justice is remarkably silent on the question of gender which may have something to do with geographies of gender (where environmental discrimination is written on the microgeographies of the body and household rather than more visible neighbourhoods and communities) and the gendered nature of the environmental justice movement itself (Buckingham, 2007). The following section will explore the contribution ecofeminism makes to critiques of the Western environmental world view, and explore its greater propensity to include ecological justice in calls for reforming the relationship between society and nature.

Ecofeminism

Ecofeminism is included here for a number of reasons: firstly, gender is habitually neglected as a way of understanding human–environment relationships by academics (other than feminist scholars), campaigners and policymakers; secondly, partly as a response to this, exploring the relationship between gender and environment is the major research concern of the author; thirdly, a focus on gender provides an example of a relationship – patriarchy – other than capitalism/neo-liberalism which structures social relations. Moreover, ecofeminism answers Schlosberg's critique of environmental justice: that it fails to consider ecological justice alongside environmental justice. As a strategy, or campaign, ecofeminism requires revisions to both how society relates internally, and how it relates to its environment, seeing these relationships as interlinked and impossible to be addressed in isolation, without negatively harming the other. Originally, the two strands of thought which made up ecofeminism came from essentialist and constructivist traditions. Essentialism, which argued that women have innate qualities which predispose them to be more sympathetic with the environment, has not gained widespread support and has caused many to question the appropriateness of ecofeminism itself. It could be argued that there are biological essentials which distinguish women from men, particularly with regard to their capacity to bear children. Women are more active, and better represented, in grass-roots environmental campaigning than in decision-making positions (see

Table 2.1   Percentage of women in decision-making positions in Europe and the USA

Institution	% Women
Members of the European Parliament[1]	28%
Members of national parliament, EU average[1]	23%
Members of EU upper-house legislature[1]	21%
US Senators, 2005	14%
Board members of Europe's top 200 companies, 2004[2]	8%
Executive Directors, FTSE 350 companies[2]	3%
Non-executive directors, FTSE 350 companies[2]	10%

Sources:
[1]European Commission, Employment and Social Affairs, 2004;
[2]Deloitte and Touche LLP, 2006

Table 2.1), and, when interviewed, frequently cite their transition to motherhood as a key reason for their increased environmental awareness. One woman, interviewed for research on women's environmental campaigning, explained motivations behind her campaigning against an incinerator in South Wales (she set up PAIN: parents against the incinerator 'with a bunch of mums over cups of tea') as having a baby which 'changed her life, previously not much concerned with environmental problems' (Buckingham, 2006: 86). Whether it is the biology which affects these women's attitudes, or the social factors which result from the biological act of bearing children, is open to question.

Arguments based on such transitions which may have an impact on their relationship to nature, and which could not be argued for any other social group, have not sat easily with arguments that it is women's social exclusion which places them in a particular relationship with nature. Contributory factors to this exclusion include discrimination in work which results in the average pay for women being lower than for men (a point developed in Chapter 3) and in there being an asymmetrical distribution of jobs where high status, professional jobs are dominated by men and low status, routine jobs are dominated by women. Positions of power – whether in the board room of companies, in government and elected offices and even in senior positions in environmental non-governmental organisations (ENGOs) – are heavily dominated by men, as Tables 2.1 and 2.2 demonstrate. This is likely to have an impact on how decisions are made (Bhattar, 2001). Although it is not possible to establish a causal link, it is worth noting that in countries with relatively high proportions of women as legislators, there is also a greater emphasis on sustainability, as cursory analyses of Welsh, Swedish and Norwegian policy confirms. Sweden was the first country worldwide to introduce a quota system to increase the number of women in Parliament, and was the first country to have reached gender parity in its representation. In the devolved government in Wales, 56% of its Cabinet appointments were women in 2005; it was also the only UK legislature which had a statutory 'sustainable development' policy. These unbalanced divisions

Table 2.2   Gender profile of major UK environmental organisations, 2004–2005

Organisation	Chief Executive	Chair	Board of Trustees
Council for the protection of rural England	Male	Male	Male president, all 5 vice-presidents male; all 5 national executives male
Friends of the Earth	Male	Male	9 male, 4 female
Greenpeace	Male	Female	All male
National Trust	Female	Male	9 male, 2 female
Royal Society for the Protection of Birds	Male	Male	Key positions male
WWF UK	Male	Male	11 male, 2 female

Source:
Buckingham, 2007: 76

of labour in paid work are reflected in the home, with the bulk of domestic work still being undertaken by women (cooking, cleaning, shopping and caring). Issues such as relative poverty, exposure to environmental problems and choices through paid and unpaid work, result in women experiencing environmental problems more acutely than men for socio-economic reasons.

Ecofeminist arguments have been particularly concerned with ensuring that it is societies' attitudes towards both women and nature, which must be considered in parallel and which must be taken into account if, in the words of Schlosberg (who does not consider gender or ecofeminism), the gap between environmental and ecological justice is to be bridged. Andrea Nightingale, in a post-structuralist analysis of the relationship between gender and environment argues that as long as the environment remains the focus for attention, as it appears to be within mainstream environmental debates, then, in the face of impending 'risk', 'it is difficult to make a clear argument about why we need to care if men and women have different experiences and knowledge of that risk' (Nightingale, 2006: 170). Through her work in Nepal's community forests, Nightingale proposes that 'gender' and 'environment' are mutually constituted, and that attention to the ways in which gender is 'performed' 'is crucial for understanding how environmental issues come to be environmental in the first place' (Nightingale, 2006: 172). For example, when environmental problems are considered to be more important than social inequalities, strategies taken to address these problems can make these social inequalities worse. Nightingale has noted in Nepal how the social practices of leaf litter collection are gendered; where these are now being seen as ecologically destructive this has led to changes in practice. However, without the involvement of women in this negotiation, or the involvement of men in the collection, the revised practice necessitates women in more time-consuming and physically more intensive work, which in turn reinforces gender inequality. Changing ecological practice without addressing gendered practices, then, can be problematic (Nightingale, 2006: 176–7).

At the root of any ecofeminist approach is an understanding that key social structures (that is, the power relationship between males and females) must be reformed if there are to be significant environmental gains.

Buddhism and deep ecology

Dobson's 'ecologism' referred to earlier represents such a departure from the globally dominant way of seeing the world that, to anyone born and raised in the neo-liberal capitalist West, it is likely to seem an unrealistic dream (or nightmare, depending on one's point of view). This is the power of naturalising discourse. What, then, inspires people to consider a radically different alternative? For Westerners uncomfortable with the prevailing ideology, indigenous cultures and alternative spiritual traditions have provided some of this inspiration. Arne Naess (1972) distinguishes 'deep ecology' (incidentally, where this term originates), which is interested in the intrinsic value of nature, from 'shallow ecology' which is more akin to Dobson's description of environmentalism, and which is defined by its anthropocentrism. This does not, however, answer the question of how any human being might be able to think in a way that is not human-centred. Perhaps a better way of defining the deep ecology movements is by their reference to an holistic world view; indeed, the definition of ecology is the study of inter-relationships between plants, animals and the environment in which they (we) live (Dobson, 2000: 40).

One strand of the deep ecology – or deep green – critique of current environmental practice under neo-liberalism comes from Buddhist influences, both in the West and the Global South. The work of Gary Snyder from the 1960s onwards drew together Buddhist considerations of nature as a refuge, of respect and compassion for all (human and non-human) living things, the practice of *ahimsa* – the absence of desire to kill or harm and the interconnectedness of all life, with indigenous American approaches and an emerging radical environmental movement (Kaza, 2006). Snyder himself describes the future coming together of different philosophies in a rather fine geographical metaphor which bears quoting in full as it also relates to the paradigm shifts, power of discourse and the nature of myth introduced earlier in this chapter. It is worth taking some time to understand what he is saying.

'Measured in centuries and millennia, it can be seen that philosophy is always entwined with myth as both explicator and critic and that the fundamental myth to which a people subscribe moves at glacial speed but is almost implacable. Deep myths change on something like the order of linguistic drift: the social forces of any given time can attempt to manipulate and shape language usages for a while, as the French Academy does for French, trying to stave off English loan words. Eventually languages return to their own inexplicable directions.

'The same is true of the larger outline of world philosophies. We (who stand aside) stand on the lateral moraine of the glacier eased along by Newton and Descartes. The revivified Goddess Gaia glacier is coming down another valley, from our distant pagan past, and another arm of ice is sliding in from another angle: the no-nonsense meditation view of Buddhism with its emphasis on compassion and insight in an empty universe. Someday they will probably all converge, and yet carry (like the magnificent Baltoro glacier in the Karakoram) streaks on each section that testify to their place of origin. Some historians would say that "thinkers" are behind the ideas and mythologies that people live by. I think it also goes back to maize, reindeer, squash, sweet potatoes and rice. And their songs.

'It is appropriate to feel loyalty to a given glacier; it is advisable to investigate the whole water cycle; and it is rare and marvellous to know that glaciers do not always flow and that mountains are constantly walking.' (Snyder, 1990: 60–1)

Snyder's approach stresses a holistic view of nature, and human beings' place in it, and the importance of the bioregion as the most logical unit of habitation. Reflecting on the bioregion, first introduced earlier in the chapter, Snyder relates the concept to the cultural areas of the 'major native groups of North America' (Snyder, 1990: 37).

A contemporary Buddhist monk and peace campaigner, Thich Nhat Hanh, exiled from Vietnam during the Vietnam War and now living mainly in France, has interpreted Buddhism's precepts, or 'trainings' as:

- Compassion (not to kill directly, or indirectly by complicity).
- Generosity and loving kindness (sharing time, energy and resources and a determination not to steal).
- Responsible sexual relations (not to exploit other human beings, or break up families).
- Loving speech and deep listening (not to harm by word or thought, and to listen without judgement, particularly as a basis for reconciliation).
- Good physical and mental health (particularly not to imbibe/use toxins and intoxicants which would range from pesticides and fertilisers, growth hormones, alcohol, drugs, caffeine, etc).
(Thich Nhat Hanh, 1992: 82–9)

While these precepts relate to personal behaviour, and the core of Buddhist practice is personal liberation, achieved by self-practice, it is clear that if they are followed through, they have beneficial implications for non-human nature, and require that the practitioner behaves respectfully to all nature: human and non-human. Box 2.3 replicates part of a statement drafted in 1992 as a way forward for Buddhism in Vietnam. The extract which focuses on Vietnam's natural heritage is drawn from a wider statement to the Vietnamese Government encouraging religious freedom in Vietnam.

Box 2.3   Protecting our nation's nature-heritage
(preserving our mother's body)

'As Vietnamese students of the Buddha, we make a vow to protect the wholeness of the territory of Vietnam, which means to protect the soil, the mountains, the forests, the rivers, the ocean and the air. We vow to do everything that we can do to protect the environment, to protect every species of animal and plant life in the country of Vietnam. We vow to stop the pollution and destruction of the nature-heritage of Vietnam.

As Vietnamese Buddhists we call on our compatriots, our government, and all those who are friends of Vietnam anywhere in the world to make a contribution to this task of protecting the Vietnamese environment. We expect that efforts to develop agriculture and industry, investments abroad, and the exploitation of resources will be founded on the principle of protecting our nature-heritage.

'The protection of life is a practice observed by all Buddhists. Life here means not only the life of human beings, but also the life of all animal, plant, and mineral species. The Diamond Sutra teaches that the human race cannot exist if there is destruction of the animal, plant, and mineral species. Anyone living anywhere on this planet, if they are aware of the state of our planet Earth at this present time, will look at the world and act in accord with this principle.'

(Thich Nhat Hanh, 1992: 141–2)

The sentiments expressed in Box 2.3 are particularly salient considering the damage inflicted on Vietnam and Laos from the 1950s to 1973, when the last US combat troops withdrew, during which time the use of toxic defoliants, napalm and Agent Orange, by the US caused widespread lethal pollution. Moreover, when the current expansion of biofuel production in south-east Asia described earlier in the chapter is taken into account, Vietnam is arguably under greater environmental threat now than in 1992, when this was written.

Buddhism has inspired other environmental and social movements in south and south-east Asia. The Sarvodaya Movement in Sri Lanka emanates from the Mahayana Buddhist 'principles of compassion, generosity, and personal contentment.' It seeks a 'full-scale, non-violent social revolution that will fundamentally reshape modernization both in [Sri Lanka] and throughout the developing world' (Gottleib, 2006). The movement is ethnically inclusive in a country riven with ethnic divisions between Buddhist, Tamil and Muslim communities, and challenges a government driven by development imperatives.

In Europe, Fritz Schumacher, inspired by Buddhism, advocated an holistic economy based on the principles of simplicity and non-violence, the purpose of which would be to maximise human well-being with the minimum of consumption. Such an equation both fairly distributes limited and finite resources and achieves optimum efficiency. His book *Small is Beautiful* (Schumacher, 1973) has inspired a number of influential organisations, including 'Practical Action' and the Schumacher Society which have developed practical, low technology and low cost strategies for communities in the Global South (Practical Action used to be called 'Intermediate Technology'), along with education programmes. The Schumacher Society also publishes work designed to promote alternative strategies for improving the environment, one of which is written by the founders of the social enterprise, BioRegional Development Group, and advocates a rethinking of living in bioregional communities in the age of rapid mobility and information and communications technologies (Desai and Riddlestone, 2002).

These alternative systems of thought, which have inspired and are utilised by people and groups in the Global North, have been adapted by their exposure to human rights discourses, information and communications technology and the indigenous cultures of Australia and North America, as Snyder's metaphor of the glaciers has already illustrated. The value of place and a respect for natural boundaries (for example, the bioregion, as explained earlier) is combined with the ability of modern technologies to enable protest groups to mobilise rapidly against perceived environmental damage; a development which will be examined further in Chapter 3. In such ways, then, do environmental philosophies adapt and travel: products of their time and place and forged through encounters with power, and between different cultures.

Summary

Perhaps the best way to summarise this chapter is for you to examine your own views of nature and the environment in the light of where, and with whom, you grew up, your education, the newspapers and websites you refer to and the TV programmes you watch. It is only through reflecting on how your own perspectives are created that you can effectively gain control over them and hence challenge and change them.

References

Armstrong, K. (2005) *A Short History of Myth*. Edinburgh: Canongate.

Bhattar, G. (2001) 'Of geese and ganders: mainstreaming gender in the context of sustainable human development' *Journal of Gender Studies*, 10(1): 17–32.

Bookchin, M. (1990) 'Social ecology', in L.M. Benton and J.R. Short, *Environmental Discourse and Practice: A Reader*. Oxford: Blackwell.

Buckingham, S. (2006) 'Environmental action and women's citizenship', in S. Buckingham and G. Lievesley, *In the Hands of Women: Paradigms of Citizenship*. Manchester: Manchester University Press.

Buckingham, S. (2007) 'Microgeographies and microruptures: the politics of gender in the theory and practice of sustainability', in R. Krueger and D. Gibbs, *The Sustainable Development Paradox*. New York: Guilford Press.

Bullard, R.D. (1999) 'Dismantling environmental racism in the USA', *Local Environment*, 4(1): 5–19.

Burgess, J., Clark, J. and Harrison, C. (2000) 'Culture, communication, and the information problem in contingent valuation surveys: a case study of a Wildlife Enhancement Scheme', *Environment and Planning C: Government and Policy*, 18(5): 505–24.

Cafard, M. (1989) 'The Surre(gion)alist Manifesto, *in Mesechabe*, 4(5): 22–4, 32–5.

Castree, N. (2001) *Social Nature: Theory, Practice and Politics*. Oxford: Blackwell.

Checker, M. (2005) *Polluted Promises: environmental racism and the search for justice in a southern town*. New York: New York University Press.

Chrysaviggis, J. (2006) 'The earth as sacrament: insights from orthodox Christian theology and spirituality', in Gottlieb, R. (ed.), *The Oxford Handbook of Religion and Ecology*. Oxford: Oxford University Press.

Deloitte and Touche LLP (2006) 'Executive directors decline in number as FTSE boards shrink to meet Higgs code' at www.deloitte.com/dtt/press_release/0,1014, sid%253D2992%252cid%253D130907,00.html (accessed 15 October 2006).

Desai, P. and Riddlestone, S. (2002) *BioRegional Solutions for Living on One Planet*. Dartington, Devon: The Schumacher Society/Green Books.

Dobson, A. (2000) *Green Political Thought*. 3rd edn. London: Routledge.

Dunnion, K. and Scandrett, E. (2003) 'The campaign for environmental justice in Scotland as a response to poverty in a northern nation' in J. Agyeman, J.D. Bullard and B. Evans. *Just Sustainabilities: Development in an Unequal World*. London: Earthscan.

ESRC Global Environmental Change Programme (2001) 'Environmental justice: rights and means to a healthy environment for all, special briefing no.7., University of Sussex.

Flint, C. and Taylor, P. (2007) *Political Geography: World Economy, Nation-state and Locality*. Harlow, Essex: Pearson, Prentice Hall.

Gandhi, M. (1993) *Gandhi, an Autobiography: The Story of My Experiments with Truth*. Boston, MA: Beacon Press.

Giddens, A. (1994) *Beyond Left and Right*. Cambridge: Polity Press.

Gottlieb, R. (2006) 'Religious environmentalism in action', in R. Gottleib (ed.), *The Oxford Handbook of Religion and Ecology*. Oxford: Oxford University Press.

Hajer, M.A. (1995) *The Politics of Environmental Discourse*. Oxford: Oxford University Press.

Hart, J. (2006) '*Catholicism*', in R. Gottlieb (ed.), *The Oxford Handbook of Religion and Ecology*. Oxford: Oxford University Press.

Hawken, P., Lovins, A. and Lovins, L.H. (1999) *Natural Capitalism: Creating the Next Industrial Revolution*. London: Earthscan.

Hobsbawm, E. (1977) *The Age of Revolution, 1789–1848*. London: Abacus.

Kaza, S. (2006) 'The greening of Buddhism: promise and perils', in R.S. Gottlieb (ed.), *The Oxford Handbook of Religion and Ecology*. Oxford: Oxford University Press. pp. 184–206.

Klein, N. (2001) *No Logo*. London: Flamingo.

Kuhn, T.S. (1970) *The Structure of Scientific Revolutions*. 2nd edn. London: University of Chicago Press.

Kurtz, H. (2003) 'Scale frames and counter-scale frames: constructing the problem of environmental injustice', *Political Geography*, 22: 887–916.

Mason, M. (1999) *Environmental Democracy*. London: Earthscan.

Maynard, R. (2007) 'Against the grain', *The Ecologist*, March 2007, pp. 28–32.

Myers, N. and Simon, J. (1994) *Scarcity or Abundance? A Debate on the Environment*. New York: W.W. Norton.

Naess, A. (1972) 'Self-realization: an ecological approach to being in the world', in G. Sessions (ed.), *Deep Ecology for the 21st Century: Readings on the Philosophy and Practice of the New Environmentalism*, 1st edn. Boston, MA: Shambhala.

Nhat Hanh, T. (1992) *Touching Peace: Practising the art of Mindful Living*. Berkeley, CA: Parallax Press.

Nightingale, A. (2006) 'The nature of gender: work, gender and environment.' *Environment and Planning D: Society and Space*, 24: 165–85.

O'Tuathail, G. and Agnew, J. (1998) 'Geopolitics and Discourse: practical geopolitical reasoning in American foreign policy' in G. O'Tuathail, S. Dalby and P. Routledge (eds) (1998) *The Geopolitics Reader*. London: Routledge.

Pakenham, T. (1991) *The Scramble for Africa*. London: Abacus.

Pepper, D. (1996) *Modern Environmentalism*. London: Routledge.

Schlosberg, D. (2007) *Defining Environmental Justice, Theories, Movements and Nature*. Oxford: Oxford University Press.

Schumacher, F. (1973) *Small is Beautiful*. London: Blond and Briggs.

Snyder, G. (1990) *The Practice of the Wild*. New York: North Point Press.

Swyngedouw, E. (1999) 'Modernity and hybridity', *Annals of the Association of American Geographers*, 89: 443–65.

Tierney, J. (1990) 'Betting the Planet', *The Guardian*, 28 December.

Woodin, M. and Lucas, C. (2004) *Green Alternatives to Globalisation*. London: Pluto Press.

Young, B. (2006) *Farmers Weekly*, October 2006.

Appendix 2.1 – Answer to Simon–Ehrlich argument

In 1990, Ehrlich 'mailed Simon a sheet of calculations about metal prices – along with a cheque for $567.07. . . . Each of the metals chosen by Ehrlich's group, when adjusted for inflation since 1980, had declined in price. The drop was so sharp, in fact, that Simon would have come out slightly ahead overall even without the inflation adjustment called for in the bet. Prices fell for the same cornucopian reasons they had fallen in previous decades – entrepreneurship and continuing technological improvements. Prospectors found new lodes, such as the nickel mines around the world that ended a Canadian company's near monopoly of the market. Thanks to computers, new machines and new chemical processes, there were more efficient ways to extract and refine the ores for chrome and other metals. For many uses, the metals were replaced by cheaper materials, notably plastics, which became less expensive as the price of oil declined.'

John Tierney, 1990

3 The rise of environmental politics and the environmental movement

Susan Buckingham

Learning outcomes

Knowledge and understanding of:

○ the relationship between environmental non-governmental organisations (NGOs) and formal political institutions and parties at a range of scales

○ the ways in which the environmental movement has grown and changed over the past 100 years.

Critical awareness and evaluation of:

○ the degree to which environmental campaigners can achieve environmental change

○ how representative (or not) environmental NGOs are of the society in which each exists

○ how any position taken on the environment is a political position

Introduction: voting on the environment

In November 2004, 64 per cent of US citizens eligible to vote, voted in the presidential elections (US Census Bureau, 2006). Fifty-one per cent of these (roughly 62 million, or just over 32 per cent of those eligible) voted for a President who had withdrawn his country from the global climate change negotiations, while only 1 per cent voted for a candidate, Ralph Nader, who stood on an environmental platform. Nader had represented the Green Party in the USA in the 2000 election, when he polled 2.7 per cent of the vote, but stood as an independent in 2004 (CNN, 2006; Green Party of the USA, 2006). In May 2005, 61.4 per cent of the British electorate voted in elections in which the only explicitly environmental party polled 1% of the vote. As the UK uses a 'first past the post' polling method in national elections (see Box 3.1), and because the green vote is geographically dispersed, none of the votes polled were converted into seats

Table 3.1   European Union member states with Green Party members of the European Parliament, 6th parliamentary term, 2004–2009

Country	Green Group*	Total
Austria	2	18
Belgium	2	24
Denmark	1	14
Finland	1	14
France	6	78
Germany	13	99
Italy	2	78
Latvia	1	9
Luxembourg	1	6
Spain	3	54
Sweden	1	19
The Netherlands	4	27
United Kingdom	5	78
Total	**42**	**732**

Source: Europarl (European Parliament, 2004)
*Includes MEPs belonging to the European Free Alliance

won. The UK Green Party's performance at the 2004 European elections was marginally better in which two Green Party candidates were elected as Members of the European Parliament (the remaining 3 included in Table 3.1 being other parties grouped under the 'Green' umbrella in European Parliament statistics). These joined a further 40 Green Party and European Free Alliance MEPs, the majority of whom came from the older members of the European Union (2004 was the year of significant enlargement of the EU in which the Union grew from 15 to 25 member states). From the distribution of Green MEPs in Europe demonstrated in Table 3.1, it is interesting to speculate on the reasons that the countries listed have returned Green Party MEPs.

Although people's and countries' attitudes towards environmental issues will be one factor in determining the relative number of Green representatives, there are other structural factors involved such as voting systems, as Box 3.1 shows.

Box 3.1   Voting systems

The UK has been using a form of proportional representation (PR) for European, London Assembly and devolved national elections (to the Scottish, Welsh and Northern Ireland Assemblies) since 1999, and this has probably had an impact on the presence of Green representatives. For example, in addition to the MEPs referred to above, the Scottish Parliament returned seven Green Party members (MSPs) in the 2003 elections. These are elected through the non-geographical 'list' rather than the constituency – the combination of list and constituency enables proportional representation not possible through constituency alone. Similarly, in London, two out of eleven list assembly members are from the Green Party, while none were elected from the constituencies. By offering voters a more nuanced choice of candidate (where they can rank candidates by preference rather than having to select only one), PR is generally considered to favour smaller, less mainstream, parties.

Sources: http://www.london.gov.uk/assembly/lams_facts_cont.jsp;
http://www.scottish.parliament.uk/msp/index.htm

By 2007, membership of environmental groups had far outnumbered membership of political parties. The combined membership of the three main British political parties was around 700,000, while the combined membership of the fifteen environmental groups challenging the construction of the third runway at Heathrow Airport (including some of the UK's largest environmental membership organisations: Friends of the Earth, Greenpeace, the National Trust, the Royal Society for the Protection of Birds and the Council for the Protection for Rural England, along with smaller, more specialist groups such as Plane Stupid, Airport Watch, HACAN Clear Skies and the Aviation Environment Federation) was estimated at around 5 million (BBC, 2007). While there will be a degree of overlap of members between these various groups, this difference is sufficiently great to suggest that there is a relatively high level of concern for the state of the environment which people do not feel is adequately reflected in the concerns of mainstream political parties. It also raises the issue of how we view politics, which encompasses a wide range of activities of which voting, which has been reviewed in some detail above, is but one small and sporadic part. Raymond Williams, the cultural historian, has described political activity much more broadly, and in a way that contributes to an understanding of grass-roots, personal and other non-government-originated action:

'It is quite a common response among many serious people now: that politics is a superficial business, it is just the in and out of competing parties, the old Left-Right see-saw . . . as anyone knows who's knocked around in politics, "no politics" is also politics and having no political position is a form of political position and often a very effective one' (Williams, 1982).

Taking this most inclusive view, environmental politics can include political action as diverse as that coordinated by groups like Reclaim the Streets, The Land is Ours and Critical Mass, whose direct action campaigns will be discussed towards the end of this chapter, to more sedate pressure groups who campaign against road expansion and peri-urban development to now well established groups such as Friends of the Earth, Greenpeace and WWF who employ a range of strategies including campaigning, direct action and political lobbying. It can also include personal lifestyle and consumption decisions, such as individuals and households growing some of their own food, opting for public transport, cycling and walking over car ownership, buying fairly traded and organically produced food and other products, and a range of other choices made for ideological reasons. From this, it can quite clearly be seen how environmental politics and the environmentalism which underpins it is amorphous and often inconsistent, allowing, as it does, different groups and people to support different elements and policies at different times.

While there is no coherent or agreed unity of thought or purpose, the environmental movement does highlight issues which people are dissatisfied with, and symbolic events that are highly newsworthy. The very nature of this both divides people who previously might have found common alliance in their class, and brings together people from very different backgrounds. One of the author's enduring memories of the campaign against the M11 motorway extension in north-east London is the bonding of a group of 'New Age' protesters who had built temporary homes in some trees in the path of the proposed motorway, winching up fellow campaigners from the Women's Institute for whom this was a very novel strategy. Such a spectrum of participation results in a wide range of political approaches to environmental issues, both operating within structured institutional forms and the more anarchic. Environmentalism can be conservative (as in wanting to conserve nature or particular practices) and against change, or it can be radical – expressing dissatisfaction with the status quo and demanding change, as Chapter 2 has already explained. Both can be provoked by widespread alienation from traditional forms of political organisation and power, and by feelings of powerlessness.

Anthony Giddens argued in the 1990s for a reconceptualisation of the political landscape away from the red-socialist/blue-conservative division to acknowledge the importance of cross-class issues such as environment (Giddens, 1994). Indeed, in Giddens' conception, 'saving the environment' is 'a gloss for the problem of how we should cope with the double dissolution of tradition and nature . . . for the environment is no longer nature and traditions have to be decided about, rather than taken for granted' (1994: 247). While left and right may express different elements of environmental concern to service their particular political interests, Giddens optimistically suggests that a set of 'universal values', including environmental, are being forged from 'global cosmopolitanism' in and through which we recognise shared values. States in the West are now obliged, through 'freedom of information' legislation, to make information more transparent and widely available. Widespread education in the West, which has created almost universal levels of literacy and significant proportions of school leavers entering higher education, enables citizens to assess scientific information and technological choice to make moral decisions about how society should use the environment.

With all of this in mind, this chapter will now consider how the environmental movement has developed from the late 19th to the early 21st centuries, both as a result of the changing nature of society, and as a response to how this society is impacting on the environment. In reality, of course, there is no one environmental movement, as the above paragraphs reinforce – it is made up of many different interests, which occasionally combine to fight individual issues. However, a number of commentators have observed sufficient coherence among these interests to class 'the environmental movement' as a distinct social movement (notably Manuel Castells, one of the leading theoreticians of social movements – see Box 3.2 – and technology, which will be discussed later) and this practice will be followed here.

Box 3.2   Social movements

Social movements have been defined by Manuel Castells as 'purposive collective actions whose outcome, in victory as in defeat, transform the values and institutions of society' (p. 3).

They are characterised by three qualities:

1. They are what they say they are. Their practices are their self-definition.
2. They may be socially conservative, socially revolutionary, both or none.
3. They can be defined by their identity, adversary, vision and/or social goal.

While this may seem confusing or contradictory, what Castells is arguing is that as an academic, he is interested in explaining and understanding social movements, not by imposing on them some arbitrary, theoretical definition. Clearly, by accepting social movements' own self-definitions, he accepts that any analysis on this basis is likely to be chaotic (rather like the real world). The social movements he studies for his book *The Power of Identity*, from which these ideas and quotes are drawn, include the feminist movement, the gay/lesbian movement and the movement against globalisation, as well as the environmental movement.

Castells uses precisely this diversity – what he terms 'this cacophony of . . . multiple voices' – to characterise 'environmentalism as a new form of decentralized, multiform network oriented pervasive social movement (1997: 69 and 112). He also argues that, taken as a whole, the environmental movement is 'arguably the most comprehensive, influential movement of our time' (p. 69). Contemporarily, writing from a southern perspective, Shiv Visvanathan (1991) argues that groups campaigning on environmental issues are 'creating a new dance of politics', which challenges international politics and to which we will return later in the chapter.

The history of the environmental movement in the USA and Europe

In both Europe and North America, environmental concern has followed the two trajectories of conservation and public health. In one sense, these are linked in that the spread of industrialisation, whether through factories or transportation networks, was seen to threaten wilderness (in North America) and rural landscapes. Substantial and rapid rural-to-urban migrations in the 19th century had put considerable strain on what little urban infrastructure existed, and the lack of access to clean drinking water and sewerage, combined with poor housing conditions and overcrowding, led to disease. As urban and industrial areas expanded, their reach was impacting on the surrounding countryside and conservationists in the USA and UK were moved to seek legal protection for the countryside. The first national park – Yellowstone – was established in the USA in 1872, which provided a model for parks elsewhere. The US environmental movement (including the national parks and the first environmental organisation 'The Sierra Club' formed in 1892) owes a great deal to George Perkins Marsh who published 'Man and Nature' in 1864 which has inspired generations of environmentalists. He is acknowledged as the first American environmentalist in the sense that he raised concerns about the destructive impact of human activity on the environment, and argued for development to be assessed for its potential disruption of nature (Lowenthal, 2003). Perkins Marsh was also linked to the conservation-minded New England transcendentalists through his cousin. Transcendentalism brought together poets, intellectuals, environmentalists and political activitists, such as Walt Whitman, Ralph Waldo Emerson and Henry Thoreau who expressed '. . . the first outcry of the heart against the material pressures of a business civilization' (Miller, 1981: ix). As a brief aside, at this point, Henry Thoreau is a particularly interesting writer for the wider purposes of this chapter, as he was engaged not only in the transcendentalist movement, but was also a proponent of direct action – writing 'On the duty of civil disobedience' in 1848 to justify his refusal to pay taxes to a government which both sanctioned and thrived on slavery. On his imprisonment for this offence, he countered: 'Under a government which imprisons any unjustly, the true place for a just man is also a prison' (Thoreau, 1960: 230). As the examples of Arundhati Roy and Ken Saro Wiwa drawn later in this chapter show, the profile of an environmental issue is often significantly raised by the imprisonment of its antagonists.

In 19th-century Britain, the conservation of countryside was also much inspired by artists and writers, chief among them William Wordsworth. Although the first national park was not founded until 1951 (The Peak District National Park), its heritage can be found both in the American environmental movement described above, and in the founding of the National Trust in 1895 by Canon Hardwicke Rawnsley, Sir Robert Hunter and Octavia Hill, and now one of the UK's largest environmental membership organisations.

Were it not for Hill's concern to link the conservation movement with the public health movement – Hill was also a leading campaigner for decent workers' housing and added her weight to the early new towns movement by raising support for Hampstead Garden Suburb in London – such a call on behalf of the National Trust could lead to charges of elitism. Certainly the rise of nationalist movements in Europe (and especially in Germany) pressed vernacular 'nature' to their cause. Nazi Germany linked the (German) people with their blood and soil in a uniquely German concept of *lebensraum* (which signified an organic 'living space' required by the country to enable its people to thrive). Periodic failures in the economy across Europe frequently saw a shift in emphasis to a focus on the past, which included landscapes unscarred by commerce or industry. In the UK, environmental destruction at home was compared to its main industrial and military rival at the time – Germany and its protection of its traditional heritage – and this gave a political spur to environmental protection in Britain.

Within the UK context, David Pepper (1996) conceptualises this period as the first phase of environmental concern spanning 1880 to 1900. Phase 2, he suggests, is characterised by the development of the garden city movement (of which Hampstead Garden Suburb, referred to above, was a part), and more conservation activity through the foundation, in the 1920s, of groups such as the Council for the Protection of Rural England and the Society for the Preservation of Birds (now the Royal Society for the Protection of Birds, and the environmental organisation with the highest membership in the UK). This period also saw the rise of the campaign for public access to open land and the formation of the Ramblers' Association in 1935. A spectacular piece of direct action through the mass trespass of Kinder Scout in the Derbyshire Peak District in 1932, organised by the British Workers Sports Federation highlighted the extent of land held in a very few private hands and the denial of its access to the industrial working classes of neighbouring cities such as Sheffield and Manchester. This is an early example of 'environmental citizenship' – or the claiming of environment goods as a right, rather than as a concession or act of generosity by the rich. This does, in fact, have a longer, if sporadic tradition, driven by poverty and desperation. The 'Diggers' movement of the 1600s was undertaken by the landless poor, deprived of their right to grow their own food because of the enclosure acts which privatised the commons. Their protest was to plant vegetables on St George's Hill in Surrey and is one which has inspired the contemporary land rights movement 'The Land is Ours' in the UK.

From the 1950s, Pepper (1996) argues that we are in Phase 3 of environmental concern characterised by 'second generation' issues which cross national boundaries through different media (soil, water, air). By second generation, Pepper means that their consequences are not always immediately evident, either because they are subtle or delayed. Such issues include acid rain, DDT and the greenhouse effect. The 1950s, of course, was also dominated by the fear of nuclear attack and a popular political movement for nuclear disarmament – the Campaign for Nuclear Disarmament (CND) – was one of the precursors of the current environmental movement. Looking back over these periods, it is noticeable that the sudden growth of new environmental groups – in the 1890s, the late 1920s and the late 1960s – tends to occur towards the end of periods of rapid economic growth when societies are evaluating the cost of this.

Concerns about nuclear war, and, since the 1960s, chemical pollution, global warming, animal rights and so on, have undoubtedly led to a widespread increase in environmental interest, which partly explained the seven-fold increase in environmental group membership in the USA from 1960 to 1969. The number of groups, too, expanded, with the setting up of Greenpeace (1970), Friends of the Earth (1970), and WWF. A number of seminal books on environmental issues were published in the 1960s and early 1970s which propelled environmental concerns much higher on the popular political (if not party political) agenda: *Only One Earth* (Ward and Dubos, 1972); *Silent Spring* – the book in which Rachel Carson, the biochemist, exposed the link between DDT and environmental and health damage (Carson, 1965); *The Limits to Growth* (Meadows, 1972), in which

The Club of Rome (an alliance of scientists and computer modellers) predicted the running out of key resources, and the Earth's ability to absorb waste products by 2100; and Paul Ehrlich's *The Population Bomb* which focused on the need to restrict population growth if the world was not to reach the limits to growth. While the latter two books had, by the late 20th century, been effectively challenged on a number of fronts, the combined impact of these publications on galvanising public concern about the environment was significant.

It was not only the growth of the environmental issues and their reporting, however, which sparked the environmental movement. The 1960s was a time of huge expansion in higher education in Europe, North America and Australasia. The 'baby boom' generation, born into peacetime and coming of age in a period of economic growth and material prosperity, were much more likely than their parents or grandparents to move away to universities which, for students and a relatively young and rapidly expanded staff, provided a platform for the major radical protests of the decade: the French uprisings of 1968 and the anti-Vietnam protests of the late 1960s and early 1970s. Such radicalism inspired groups such as Greenpeace and Friends of the Earth, which were quick to mobilise the developing media advantages. Increasingly, TV was transmitting images instantaneously which both appalled and inspired the new generation of environmental protesters. Greenpeace, meanwhile, became adept at using the media in their campaigns – creating images with their boat *Rainbow Warrior* protesting against nuclear testing in the South Pacific, as well as using reported disasters to galvanise support and action. For the UK, one of the seminal environmental disasters of the 1960s was the oil spill from the tanker Torrey Canyon which broke up off the coast of Cornwall. Images of this have been repeated with subsequent oil spills, most dramatically, from the Exxon Valdez, an oil tanker which broke up off the Alaskan coast in 1989. As Chapter 4 will later discuss, Marilyn Waring has documented this event as an example of the irony by which significant environmental damage can translate positively in national economic accounting procedures, such as Gross and National Domestic Product (through, for example, clean up, legal and insurance transactions). Table 3.2 illustrates the damage wrought by the Exxon Valdez and itemises other major oil spills since the Torrey Canyon in 1967.

An irony emerges from this: in many respects, the 1960s represented a period of material security in which leisure time and economic wealth were increasing, and yet the environmental destruction on which this prosperity was built was causing concern. In addition to the books referred to above, which identified a range of key problems, as the authors saw them, others were published as attempts to offer an alternative and have now become seminal works. An example of this, already mentioned in Chapter 2, is Schumacher's *Small is Beautiful* (Schumacher, 1974) which proposed small-scale production as a route to self-actualisation to lift people through a hierarchy of need to altruism. His principles have inspired organisations dedicated to teaching and demonstration projects, such as Dartington Hall, Schumacher College, and Practical Action. The founders of these organisations, and participants in them, are examples of environmental political action in the 20th century. The next section considers the position of the environmental movement at the beginning of the 21st century.

Table 3.2   Major oil spills

Year	Ship	Location	Oil Spilt (tonnes)	Damage Estimated
1967	Torrey Canyon	off Cornish coast	120,000	70 miles o beaches affected by oil/detergent; tens of thousands of seabirds killed
1978	Amoco Cadiz	off west coast of France	220,000	400 km of beaches affected; 300,000 seabirds killed
1979	Ixtoc-1 oil rig	Gulf of Mexico	500,000	Atmospheric pollution from burning oil; shrimp nurseries, mangoves, seabirds affected
1989	Exxon Valdez	off Alaskan coast	38,000	1000 km of coastline contaminated; residual contamination in 1995
1993	Brear	Shetland Isles, Scotland	84,700	£45 million paid in compensation
1996	Sea Empress	South Welsh coast	72,000	200 km of coastline affected; several thousand seabirds killed
1999	Erika	Brittany, France	20,000	350 km of coast-line affected; a minimum of 100,000 seabirds killed
2002	Prestige	Galician coast, Spain	63,000	WWF predict marine life polluted for 10 years

Sources: BBC, 2007; Cedre, 2007

The contemporary environmental movement

Since the 1960s and early 1970s, the environmental movement has expanded, become more complex and fragmented. The groups themselves have changed and in some cases have become more institutionalised. New groups have broken away, frustrated at what they see as 'selling out' or not covering issues seen as important. Chris Rootes (1999) suggests that environmental groups find themselves in a 'strategic dilemma', caught between 'the radical' (he styles this 'the conscience of the movement'; it is the vanguard of environmentalism, identifying issues with which governments and often communities aren't yet ready to address, and which uses strategies bordering on the illegal to raise their profile) and 'the institutional'. The latter has lured many groups and individuals with its access to political power and decision making, a process which Neil Carter has described as 'incorporation' (2001). Former directors of Greenpeace and Friends of the Earth left those organisations to become green business consultants (e.g. David Melchett formerly of Greenpeace) and chairs of government bodies. Jonathan Porrit has recently been knighted for his services to the environment and heads up the UK Commission for Sustainable Development. In a recent book, Porritt has argued that capitalism is here to stay and that environmental campaigners should accept this and use business as a tool to make progress on environmental issues (Porritt, 2005).

In some cases, individual groups oscillate between the radical and institutional – Greenpeace commissions scientific research, lobbies national governments, the EU, the UN, but conscious of the restrictions this puts on its activities, has attempted to recapture some of its old campaigning past by shedding central office staff and involving grass-roots members more. The radical edge of environmentalism is now captured by less structured groups, who use the Internet and electronic communication to generate unpredictable and often highly visible activity through a network of what Naomi Klein (2001) has described as 'webs' and 'spokes'. Castells argues that 'more than any other social force', the environmental movement has been 'able to best adapt to the conditions of communication and mobilisation in the new technological paradigm' (1997: 128).

Some environmental groups operate in a way similar to businesses, in hierarchical structures in which ethnic minorities and women are not well represented. In the early 1990s, Joni Seager identified that around 30% of the senior personnel of environmental NGOs were women, with an extremely low representation of minority ethnic groups (varying between 0.4% of the Sierra Club's total staff to 12% of Friends of the Earth's staff, compared to 20% of the US population at the time registered as an ethnic minority; Seager, 1993). In 2004, 43% of the environmental groups which contribute to the European Civil Society Contact Group were headed up by women, and only 24% of their decision-making bodies comprised women (European Parliament, 2004). If this is the case in groups committed to gender parity (a sine qua non of membership of the ECSCG), then the situation elsewhere is likely to be bleak for women as Chapter 2 has already noted in its discussion of ecofeminism.

Largely as a result of the masculinism of environmental groups operating in the UK in the 1980s, a staff member at Friends of the Earth, Bernadette Vallely, resigned to set up The Women's Environmental Network (WEN), which has been operating since 1988 to campaign on issues of direct concern to women which, WEN believes, are still largely neglected by mainstream environmental groups (see Box 3.3). Similarly, the Black Environmental Network (BEN) was founded in 1988 to counter stereotypes of the black and ethnic minority community as urban and, as a consequence, out of place in rural landscapes. The black broadcaster and chair of the Race Relations Board, Trevor Philips, has recently pointed out the deep irony of this stereotype, given the rural and farming backgrounds of many immigrants to the UK from South Asia, Africa and the Caribbean, since the 1950s. While the Black Environmental Network has gone on to work on broader environmental justice issues, which we will consider shortly, one of the co-founders of BEN, Julian Agyeman, recently wrote a book demonstrating the continuing alienation of black and ethnic minority people not only from rural landscapes, but from the environmental movement more widely (Neal and Agyeman, 2006).

Box 3.3   The Women's Environmental Network

Photo courtesy of Women's Environmental Network

This photograph depicts a performance artist, 'Monique Toxique', commissioned by WEN to raise the profile of environmental causes of breast cancer at a WEN lobby of Parliament. The discomfort registered in the male passer-by, stopped by Monique Toxique, reinforces the sexualised taboo around women's breasts which, WEN argues, has marginalised this environmental health issue for years. WEN and the UK's public sector trades' union, UNISON, launched the campaign: 'The Big See' in 2005 which seeks to raise awareness of the links between artificial chemicals and breast cancer (Women's Environmental Network, 2007).

For some individuals and communities, environmental impact is perceived as a practical misuse of resources and habitats which we have no right to inflict on other human communities, other species and other – future – generations. This element of the environmental movement is frequently depicted as socially liberal and from the 'chattering class' (well educated, middle-class, and with sufficient income to free them from pressing concerns about livelihood). While not a 'NIMBY' ('not in my backyard') movement, which presses for the relocation of

Table 3.3   Percentage of respondents 'very worried' by selected environmental issues, 1998 (2002 in bold)

Environmental Concern	Male	Female	18–24	45–64	Degree	No Qualifications	Social Class I	Social Class IV/V
Global warming	34 **43**	37 **49**	41 **42**	37 **48**	31 **44**	28 **47**	44	33
Natural resource depletion	21	24	20	26	22	25	20	24
Household waste disposal	17 **27**	26 **38**	21 **26**	26 **34**	15 **26**	23 **40**	12	24
Radioactive waste	59	62	59	66	53	60	45	68
Toxic/ hazardous waste	56 **62**	63 **70**	46 **56**	70 **72**	53 **59**	62 **70**	46	66
Traffic exhausts	45 **48**	51 **57**	40 **47**	55 **55**	46 **46**	51 **59**	42	49
Ozone depletion	45 **46**	47 **53**	51 **46**	49 **54**	43 **48**	42 **48**	53	49
Use of insecticides, pesticides, fertilisers, chemicals	43 **37**	49 **48**	35 **25**	54 **50**	43 **38**	48 **51**	38	45
Tropical forest loss	46 **49**	42 **46**	52 **40**	48 **54**	45 **52**	39 **48**	49	40
Drinking water quality	35	42	39	37	32	43	21	42

Source: DEFRA, 2002; HMG, 1998

environmentally noxious activities away from protesters' own property and neighbourhood, this element of the environmental movement is seen as privileged. Protagonists have a measure of social and economic well-being (such as a salaried job which may enable flexible working hours) which provides them with the time and resources to devote to environmental concerns without undue sacrifice. From Table 3.3 above, it can be seen that, while those people without qualifications and classified in the lowest social class appear to be more worried about most environmental issues, those with degrees and from the highest social class appear to be more worried than others about issues with less immediate and daily impact (note in particular, rows showing responses for global warming, ozone depletion and tropical forest loss). If this is compared to data in Table 3.4,

Table 3.4 Action people claim to take with regard to environmental problems – percentage of respondents, 2002

Social Environmental Action	Male (%)	Female (%)	18–24 (%)	45–64 (%)	Degree (%)	No Qualification (%)	Social Class I (%)	Class IV/V (%)
Take paper for recycling	52	54	29	59	62	51	61	45
Make compost	22	18	8	26	28	20	26	15
Cut down use of car	39	38	36	38	37	38	34	39
Buy goods with less packaging	11	13	11	15	14	11	9	12
Buy organic	17	19	15	18	31	9	34	14
Reduce exposure	46	58	40	57	58	46	55	46

Source: DEFRA, 2002

however, which shows the action that members of different socio-economic groups claim to take, it is striking that concern and action are not necessarily linked. In fact, we might go as far to suggest that people are more likely to express concern about exactly those issues they feel are out of their control. This relates to Ulrich Beck's theory of a 'risk society', in which Beck also argues that Western society perceives greater risk as a result of its immediate material and health and safety needs being satisfied (Beck, 1995).

Environmental justice

In contrast to a predominantly Western, white and middle-class environmental movement, there is growing concern about the distribution of environmental problems, such as toxic waste and pollution. Increasingly visible at both the sub-national and global scale are communities which are directly and negatively impacted by environmental abuse because of their relative poverty and disadvantage. Such communities are protesting these environmental incivilities as cases of environmental injustice: where people and communities suffer disproportionate environmental problems because they are poor – or black – or powerless indigenous communities. Like the more privileged campaigners referred to earlier, these environmental justice campaigners not only oppose environmentally noxious activities in their backyard, but argue for a rethinking of the way in which society is organised so that the environment does not have to be so degraded in the first place. In the USA, this movement gathered momentum from the African-American Civil Rights movement, largely from the 1980s, and has tended to focus on the experience of non-white communities which appear to have been systematically targeted, as levels of their exposure to pollution and toxic waste is disproportionately high, even when controlling for poverty (Bullard, 1999, and Chapter 2). Box 3.4 gives an example of an environmental justice campaign against the Narmada dams in India which illustrates both the extent of environmental protest in a country in the Global South, and the global reach of economic activity, which connects investment decisions made in the economic nerve centres in the West with the lives of millions of poor, rural inhabitants of the Narmada River Valley in India. This campaign has attracted worldwide attention and has contributed significantly to the development of the global environmental movement, providing a potent emblem of economic greed and environmental injustice on a huge scale.

Box 3.4   The Narmada dams

Three thousand two hundred dams are being planned along the Narmada River in central India, altering the ecology of the entire river basin of one of India's biggest rivers. Four thousand square kilometres of natural deciduous forest will be submerged. The Narmada dams will add to the existing 3300 'Big Dams' already built in the name of development. According to the novelist, Arundhati Roy, who is a leading campaigner against the proposed Sardar Sarovar dam, the ongoing fight against this dam represents wider concerns about an entire political system and about land and resource ownership. Despite India's enthusiasm for building 'Big Dams' since its independence (Nehru claimed them to be the 'Temples of Modern India' before he rescinded this idea in favour of the small irrigation systems that were more likely to change the country; Roy, 1999) over 200 million (20%) of India's population do not have access to safe drinking water and 600 million lack basic sanitation. Roy argues that despite this phenomenal investment, there are more drought-prone areas currently than there were in 1947. That many Indians are not benefiting from this expenditure is not the only environmental injustice – displacement is a major problem:

'According to a detailed study of fifty-four large dams done by the Indian Institute of Public Administration, the *average* number of people displaced by a large dam in India is 44,182. Admittedly, 54 dams out of 3300 is not a big enough sample. But since it's all we have, let's try to do some rough arithmetic. A first draft.

'To err on the side of caution, let's halve the number of people. Or, let's err on the side of *abundant* caution and take an average of just 10,000 per one large dam. It's an improbably low figure, I know, but . . . never mind. Whip out your calculators: 3300 × 10,000 = 33,000,000.' (p. 19)

Roy goes on to protest that around 60% of the displacements of the Sardar Sarovar dam are *Adivasis* (the official Indian term which describes indigenous people) or *Dalits* (previously known as the outcaste 'untouchables'). Given that these groups represent

8% and 15% respectively of India's population, this illustrates a clear case of disproportionate environmental injustice falling to the poorest and least politically well represented groups in Indian society. As a result of this displacement, and the conditions in
which many of these now landless people find themselves (absorbed into the slums of
huge cities), Roy calls them 'nothing but refugees of an unacknowledged war' (Roy,
1999: 24).

Roy's essay, 'The Greater Common Good', continues with a litany of other negative
effects of the dam on poorer populations including deforestation, destruction of land on
which subsistence crops are grown, an increase in malaria in the dam building areas
and so on. Roy herself has been imprisoned for protesting against the dam, which
arguably secured more coverage in the West for the protests against the dams than
would otherwise have been the case.

The Narmada project is being supported by the World Bank although, as with many
development loans from the West to the Third World, India ends up paying more back in
interest repayments than it receives from the West.

Environmental justice is an increasingly important component of an environmental
movement, which gained significant momentum in the 1960s and is one example
of success achieved by an environmental campaign. In addition to forcing consideration of environmental impacts on minority Americans, environmental justice
has also entered the policy remit of the Greater London Authority, which published guidance in 2005. While race – or ethnic minority – environmental discrimination has not been specifically identified in the UK, the groundbreaking
work of Friends of the Earth Scotland (the first Friends of the Earth organisation
to explicitly address environmental justice and hence move the environmental dialogue into areas of disadvantage) has clearly established a link between poverty
and social disadvantage and environmental problems such as toxic industrial emissions (Dunnion and Scandrett, 2003; Friends of the Earth Scotland, 2000).

The one relatively disadvantaged group that is not actively considered by environmental justice campaigners, however, is women, as Chapter 2 has already
demonstrated. There are a number of reasons why women are disproportionately
negatively impacted by environmental issues compared with men, and the chapter has already explained that, as campaigning groups are themselves gendered,
these will not necessarily reveal these disadvantages or campaign against them.
Four key factors structure this gendered environmental disadvantage:

1. Women are more likely to be in poverty than men. The UN estimates that
   70 per cent of those officially defined as in poverty worldwide are women
   (United Nations, 2000). In the UK, women earn anything between 61% of
   men's hourly rate of work, if they are working part time to 81% working full
   time (Bradshaw et al., 2003; Kingsmill, 2001). In the USA, women in full-time
   work earn 76% of the equivalent male wage, with African-American women
   and Latina women earning 60% and 55% respectively (National Commission
   on Pay Equity, 2004). That many women will be heading up households in

relative poverty means that they will be disproportionately exposed to environmental injustice.

2. Women undertake the bulk of domestic chores worldwide, including cooking, cleaning, shopping and subsistence food growing, as well as childcare and care of other dependants. This exposes women to a range of pollutants in the home for which there is inadequate testing and legislation.

3. That chemical exposure traditionally has been tested and monitored on a fit male body means that their effects on women's bodies (and incidentally children and elderly frail people) are unknown. The new regulatory framework for the Registration, Evaluation and Authorisation of Chemicals that has passed into law in the European Union has identified that 'derived no-effect levels' of chemicals may need to be identified separately for different populations. However, women – including pregnant women – are rarely mentioned in the REACH documentation or the European Environmental Assessment Report (CEC, 2004; EEA, 2003).

4. Women are less likely to be in positions of power in governments worldwide, business and industry. Consequently, they are less likely to have an impact on decisions which materially affect environmental and demographic well-being (Buckingham et al., 2005b; Buckingham, 2006 and see Chapter 2).

Some of the most active work being done to redress gender inequality in environmental decision making is at the international level, as this chapter will shortly address, although there is also a lot of grass-roots campaigning work being undertaken by women – an arena where women tend to be more active than men, as Box 3.3 has shown.

Third World environmentalism

The environmental groups discussed so far are, of course, heavily dominated by Western and developed world interests and there are critical voices from the Global South cautioning against the West – even a radical West – setting the global environmental agenda. There is an active environmental politics as we have seen, in India, where the Bachao Andolan opposes the Narmada dams project. India is also the home of the Chipco movement, one of the best known and emblematic environmental protests in which loggers were opposed by members of a rural forest community by hugging trees (from which the name Chipco derives – see Vandana Shiva, 1993, for more detail). Deforestation also spurred the Kenyan Green Belt movement led by Waangari Maathi, awarded a Nobel Peace prize in 2004 for her commitment to environmental well-being in a difficult political situation – the group planted trees. Environmentalism is frequently at the forefront of political protest and in Central and Eastern Europe in the 1980s, prior to the collapse of the centrally planned economies, protesters were emboldened to protest against the proposal for a series of dams planned on the Danube. Such protest carries a political cost and not all governments are receptive to environmental lobbying, however. The case of Shell's involvement in Nigeria is a good illustration of this. In 2000, Nigeria was the fifth largest oil producer of the Organization of the Petroleum Exporting Countries (OPEC), exporting one million barrels of oil a day, of which Shell Petroleum Development Company (a subsidiary of Royal Dutch Shell) produced half. Most of the oil production has been concentrated in the Niger Delta, an ecologically rich flood plain inhabited by ethnic minority groups with marginal, at best, access to political power. These groups suffer the negative environmental consequences of oil exploitation (for example, soil and water pollution, which affects the livelihoods of farming and fishing communities and atmospheric pollution from gas flaring), without gaining any of the benefits. Between 1976 and 1996, the Niger Delta Environmental Survey established that almost 2.5 million barrels of oil had been spilt in 4647 incidents. Those grass-roots opposition groups, the best known of which is MOSOP (the Movement for the Survival of Ogoni People), which have mobilised to raise the international profile of these injustices have been systematically targeted by the Nigerian Government, culminating in the execution of its spokesman, Ken Saro Wiwa in 1994 (Agbola and Alabi, 2003: 280.)

International environmental politics

The success of international institutions in achieving environmental improvements very much depends on the the viewers' vantage point. Vandana Shiva (1993) argues that in some ways the internationalisation of environmental politics has actually closed down a number of debates, particularly those of most concern to people in the Third World. The Global Environmental Facility, set up by the United Nations as a result of the UN Conference on Environment and Development (UNCED) in 1992, but administered by the World Bank, addresses, according to Shiva, only four environmental issues:

1. climate change minimisation
2. protection of biodiversity
3. pollution of international waters and
4. reducing ozone layer depletion.

In some ways, this parallels the inconsistency considered earlier, between the wealthier and more securely placed members of society registering greater concern over the more abstract issues of climate change and ozone depletion, while those with lower social status and less well off are preoccupied with environmental issues with more immediate and local effect.

Biodiversity is an interesting case to consider in this light. Initially, support for the UN Convention on Biodiversity was withheld by the USA on the grounds that US pharmaceutical companies stood to lose from having to contribute to the protection of the biodiverse habitats they exploit. Although the Convention was ratified by the US on the election of Democrat Bill Clinton to the presidency, biodiversity continues to be a resource-based discourse which does not sit easily with debates on genetically modified crops, as Chapter 6 on food discusses.

The United Nations Conference on Environment and Development (UNCED) was organised in the wake of the report 'Our Common Future', the report of the World Commission on Environment and Development (WCED). The report propelled and justified the term 'sustainable development' into common currency and, consequently, bears a responsibility for much of the subsequent debate on environmental (and social and economic) sustainability. Visvanathan questions the logic of the term 'sustainable development' (as have many others): 'Sustainability and development belong to different, almost incommensurable, worlds. We were told in catechism class that even God cannot square the circle. Sustainable development is another example of a similar exercise' (Visvanathan, 1991: 238). The WCED was, of course, not unaware of this and such a contradiction was accepted to attempt a reconciliation between the Global North on the one hand and its concerns about climate change together with industrial and economic control based on increasing consumption, and the Global South on the other hand with its concerns about resource dependency and development. Sustainable development remains a contradiction and a poorly understood concept, even as it has entered popular discourse, as the previous chapters have indicated.

These tensions which underlay the work of the WCED have continued to structure UN debates through the 1990s and 2000s and although many treaties have been negotiated and agreements signed (as Chapter 8 explains with reference to climate change), there is little evidence of many environmental problems actually being resolved. Nevertheless, it is worth considering two generic, interlinked, approaches which UNCED pioneered. UNCED was the first UN conference to reserve a specific space for non-state actors. While this space was miles away from the formal conference (making it difficult for NGO participants to meet with politicians and policy makers), and located among the favelas of Rio de Janeiro, it achieved much publicity and the three years of NGO preparation for the conference had a significant impact. This preparation took the form of 'preparatory committees' in which policies were thrashed out which could be put to the UN to be discussed by member states. This work led to Agenda 21 and to the Rio Principles which are arguably the most enduring and successful elements of UNCED (Osborn and Bigg, 1998). In these statements lies the second approach, which is the inclusion of the minorities and the disadvantaged into deciding how environmental policies should be developed and enacted. Within Agenda 21 are chapters arguing the importance of the involvement of women, children and young people, trade unions, and indigenous people. The International Council for Local Environmental Initiatives (ICLEI) also prepared for the conference by drawing up principles by which local governments should be involved in securing more sustainable environments.

It is unwise, however, to consider NGO involvement in international conferences as a universal good. This chapter has already indicated that NGOs are not necessarily representative or democratic. Visvanathan suggests that for Brundtland (and, by extension, we might include all UN conferences and policies) to seek 'a co-optation of the very groups that are creating a new dance of politics . . .' reveals a lack of understanding of the nature of such groups. In this sense, the involvement of NGOs in international politics is an attempt to co-opt them into a (in his eyes mistaken) belief that 'the expert and the World Bank can save the world.' 'What it fails to understand is that a club of experts . . . is an inadequate basis for society' (Visvanathan, 1991).

Notwithstanding criticisms, it is possible to see from the above discussion how the nation-state has become squeezed between pressure from international organisations and from NGOs, and much of state environmental policy can be seen to be responsive to pressure, rather than pioneering. This is a point that will be taken up in the summary when the points at which individuals and groups can make an impact on (environmental) policy and political structures are discussed.

The greening of governments

In the UK, it is generally accepted that national environmental policy has been created largely under twin pressure from international agreements (from the EU and UN) and from grass-roots and some local governments. Measures, such as the National Parks, Garden Cities, Public Health Acts, Green Belt and more recently animal rights, renewable energy, waste minimisation, organic and genetic-modification-free food, have all largely been in response to grass-roots political activity. Occasionally, there have been enthusiastic environmental ministers, but this is the exception rather than the rule, although there are a handful of politicians who have been consistently supportive and innovative on environmental issues (for example, Margot Wallstrom as an EU Commissioner for Environment in 1999–2004 and Gro Harlan Brundtland as Prime Minister of Norway for 10 years in 1981, 1986–1989 and 1990–1996). As a rule, the UK government has preferred industry agreements where industrial sectors self-regulate, such as for packaging. European environmental policy has largely been driven by a coalition of richer Northern member states (Germany, the Netherlands, Sweden, Denmark, Finland, Austria) who have a tradition of conservation and, post-World War II, of social democracy. The poorer states have less bargaining power, and smaller and generally less environmentally damaging economies. In Europe, the European Union has been a powerful engine for environmental improvements, and all states joining the EU must achieve predetermined environmental standards. Chapter 8 on climate change explores the role of the EU in the climate change negotiations which culminated in the passing of the Kyoto Protocol into international law on 15 February 2005.

The election of Bill Clinton in 1992, with Al Gore and his strong environmental credentials as his Vice President, gave a measure of optimism to the US environmental movement. Significant achievements included the overturning of George Bush Senior's refusal to sign the Biodiversity Treaty, protection of the Alaskan Wilderness and Pacific NE Forests, and the environmental justice measures referred to above. However, the subsequent presidency of George W. Bush has reversed, or neglected, a number of these achievements, and US environmental legislation is weaker now than in the 1990s.

As well as some impact on international agreements, particularly since 1992 which has been discussed above, ENGOs have also been credited with significant roles in international legislation such as the Convention for International Trade in Endangered Species (CITES), the Montreal Protocol on Ozone Depleting Substances and Agenda 21, although as Jordan (1998) has pointed out with regard to Montreal, the business context needs to be supportive for this to happen, as the discussion in Chapter 2 has explained. ENGOs have a much tougher time when running up against powerful industries whose business interests do not coincide with the environmental agenda, as the explanation of damage by oil exploitation on the Ogoni region of Nigeria has already shown, and which Chapter 8 on climate change points out.

As this chapter was being finalised, environmental issues had been moved up the political agenda with the election of David Cameron as the leader of the Opposition Conservative Party in the UK. Cameron declares himself personally committed to improving the environment and has made a lot of political capital from his own lifestyle decisions (such as environmentally sound technology in his home, and cycling). While his position has forced other parties into defending what green credentials they have, it is by no means certain that Cameron will carry the Conservative Party with him on this. Nevertheless this debate, and the proliferation of 'green' articles in magazines in 2006, has ensured a wider airing for environmental politics (see *Vanity Fair*, 2006 and Figure 3.1 for examples). This debate also echoes Giddens' 'third way' analysis of politics beyond left and right, which this and the previous chapter have considered earlier.

Finally, local politics has often been the bastion of environmental concerns, galvanised by people's immediate concerns about air quality, congestion, noise, and general environmental incivilities, such as dog mess and graffiti. While the Green Party has never secured a foothold in national UK politics, in 2007 it won its largest ever number of seats on local councils.

A number of local governments have developed pioneering green practices, whether it is implementing a traffic congestion charge (Greater London Authority); providing a service for reusable nappies to replace the disposables which make up 3% of the UK's landfill (West Sussex, UK; see Buckingham et al., 2005b); financially rewarding new building which achieves stringent environmental standards (Portland, Oregon); dismantling years of unsightly and ecologically damaging beach-front development (Palma, Mallorca; see Evans et al., 2005); or working with designers of zero energy homes to review planning requirements (BedZED, London, Leicester City Council; see ZEDFactory, 2007). Local groups of campaigners, too, continue to launch campaigns which achieve a mixed degree of success, depending on particular circumstances in specific places and at specific times, as this chapter has illustrated.

Summary

This chapter began with Raymond Williams' exhortation to consider political activity as any position anyone has on anything. While it has drawn a picture of a heavily structured environment in which political decisions are made, and the difficulties of achieving change, it does also signify the potential for change, given the right circumstances. Michael Mason has called this the 'political opportunity structure', referring to ways in which certain radical interests can be inserted into

Figure 3.1 *Time Out*, 'Go Green' issue. (Reproduced with permission from Time Out Group, Issue 1855, March 8–15, 2006.)

policy if the opportunity is right (Mason, 1999). At the international level, the opportunity for this was created by technological changes at a time when campaigners were calling for the control of ozone-depleting substances. At present, there is a major debate on energy power sources, as friction in the Middle East reveals the extent of the West's oil dependency, which gives leverage to renewable energy sources and energy conservation.

Most societies are conservative by nature, and resistant to change. Governments work to short time frames of electoral periods and are reluctant to engage in expensive programmes which will not manifest benefits until their term of office is over. Much of the impetus for environmental legislative or behavioural change comes from what has been loosely described as the 'environment movement', although as this chapter has shown, this is fuzzy, messy and sometimes anarchic. Despite this, however, significant changes in position have taken place on a variety of scales, most recently on the international scale, although the extent to which this filters through to the national is still questionable. Likewise, strong, imaginative local initiatives from governments, or campaigns against genetically modified food (in Europe), fur, fast foods, or in favour of animal rights, have gained enough traction to influence legislation. Alain Lipietz, academic and French Green Party politician, argues that, despite dominant structures in society which work against social and environmental change, it is possible to work at incrementally 'rupturing' these structures to achieve social and environmental gains (Lipietz, 2000). However, existing political structures are very robust and gains or changes made need vigilance to ensure they can be maintained.

References

Agbola, T. and Alabi, M. (2003) 'Political economy of petroleum resources development, environmental injustice and selective victimisation: a case study of the Niger Delta region of Nigeria', in J. Agyeman, R. Bullard and B. Evans (eds), *Just Sustainabilities, Development in an Unequal World*. London/Cambridge, Mass: Earthscan/ MIT Press.

Agyeman, J. (2000) *Environmental Justice*. London: TCPA.

Agyeman, J. Bullard, R. and Evans, B. (2003) *Just Sustainabilities: Development in an Unequal World*. London/Cambridge, Mass: Earthscan/MIT Press.

Beck, U. (1995) *Ecological Politics in an Age of Risk*. Cambridge: Polity Press.

BBC (2007) http://news.bbc.co.uk/onthisday/hi/dates/stories/march/18/newsid (accessed 7 August 2007).

Bradshaw, J., Finch, N., Kemp, P.A., Mayhew, E. and Williams, J. (2003) *Gender and Poverty in Britain*. Manchester: Equal Opportunities Commission.

Buckingham, S. (2006) 'Environmental action as a space for developing women's citizenship', in S. Buckingham and G. Lievesley (eds), *In the Hands of Women: Paradigms of Citizenship*. Manchester: Manchester University Press.

Buckingham, S., Budd, J., Lynn, H., Murphy, D. and Sutton, L. (2005b) *Why Women and the Environment?* London: Women's Environmental Network.

Buckingham, S., Reeves, D. and Batchelor, A. (2005b) 'Wasting Women: the environmental justice of including women in municipal waste management', *Local Environment,* 10: 4.

Bullard, R.D. (1999) 'Dismantling racism in the USA', *Local Environment,* 4:1: 5:9.

Carson, R. (1965) *Silent Spring*. Harmondsworth: Penguin Books.

Carter, N. (2001) *The Politics of the Environment: Ideas, Activism, Policy*. Cambridge: Cambridge University Press.

Castells, M. (1997) *The Power of Identity*. Oxford: Blackwell.

Cedre (2007) http://www.cedre.fr/index_gb.html (accessed 7 August 2007).

CE (Commission of the European Community) (2004) 'Registration, evaluation and authorisation of chemicals' http://europa.eu.int/scadplus/leg/en/lvb

CNN (2006) cnn.com/ELECTIONS/2004/pages/results (accessed 29 March 2006).

DEFRA (2002) defra.gov.uk/environment/statistics/pubatt/download/csv/pa01tbl4b.csv (accessed 19 January 2005).

Dunnion, K. and Scandrett, E. (2003) 'The campaign for environmental justice in Scotland as a response to poverty in a northern nation' in J. Agyeman, R. Bullard and B. Evans (eds), *Just Sustainabilities, Development in an Unequal World*. London/ Cambridge, Mass: Earthscan/MIT Press.

Ehrlich, P.R. (1969) *The Population Bomb.*: The Sierra Club

EEA (European Environment Agency) (2003) *Environmental Assessment Report*. Copenhagen: EEA.

European Parliament (2004) www.europarl.eu.int/ep6/owa/p_meps2.repartition?ilg=EN&iorig=home (accessed 4 October 2004).

Evans, B., Joas, M., Sundback, S. and Theobald, K. (2005) *Governing Sustainable Cities*. London: Earthscan.

Friends of the Earth Scotland (2000) *The Campaign for Environmental Justice*. Edinburgh: Friends of the Earth Scotland.

Giddens, A. (1994) *Beyond Left and Right*. Cambridge: Polity Press.

Green Party of the USA (2006) http://www.gp.org

HMG (1998) *UK Digest of National Statistics 20*. London: The Stationery Office.

Jordan, A. (1998) 'The ozone endgame: the implementation of the Montreal Protocol in the United Kingdom', *Environmental Politics*, 7(4): 23–52.

Kingsmill, D. (2001) *Kingsmill Report on Women's Employment and Pay*. London: DTI.

Klein, N. (2001) *No Logo*. London: Flamingo.

Lipietz, A. (2000) 'Political ecology and the power of marxism', *Capitalism, Nature, Socialism*, 11(1), electronic copy, unpaged.

Lowenthal, David (2003) *George Perkins Marsh: Prophet of Conservation*. Washington: University of Washington Press.

Mason, M. (1999) *Environmental Democracy*. London: Earthscan.

Miller, P. (1981) *The American Transcendentalists*. Baltimore, MD: Johns Hopkins Press.

National Commission on Pay Equity (2004) www.pay-equity.org/index.html (accessed 18 February 2005).

Neal, S. and Agyeman, J. (eds) (2006) *The New Countryside? Ethnicity, Nation and Exclusion in Contemporary Rural Britain*. Bristol: The Policy Press.

Osborn, D. and Bigg, T. (1998) *Earth Summit II: Outcomes and Analyses*. London: Earthscan.

Pepper, D. (1996) *The Roots of Modern Environmentalism*. London: Routledge.

Porritt, J. (2005) *Capitalism as if the World Matters*. London: Earthscan.

Rootes, C. (1999) 'Environmental movements: from the local to the global', *Environmental Politics*, 8 (1): 1–12.

Roy, Arundhati (1999) *The Cost of Living*. London: Flamingo Press.

Schumacher, F. (1974) *Small is Beautiful*. London: Abacus.

Seager, J. (1993) *Earth Follies: Coming to Feminist Terms with the Global Environmental Crisis*. London: Routledge.

Shiva, V. (1993) 'The Greening of Global Reach' in W. Sachs (ed) *Global Ecology: a new arena of Political Conflict*. London: Zed Books.

Thoreau, H.D. (1960) *Walden and Civil Disobedience*. New York: Signet Classics.

The United Nations (2000) *Millennium Goals*. New York: United Nations.

US Census Bureau (2006) *Voting and Registration in the Election of November 2004*. Washington: US Census Bureau.

Vanity Fair (2006) Special Green Issue, May 2006.

Visvanathan, S. (1991) 'Mrs Brundtland's disenchanted cosmos', *Alternatives*, 16 (2).

Ward, B. and Dubos, R. (1972) *Only One Earth* Harmondsworth: Penguin.

Waring, M. (1989) *Counting for Nothing: What Men Value and What Women are Worth*. Wellington, New Zealand: Allen and Unwin.

Williams, R. (1982) *Socialism and Ecology*. London: Socialist Environment and Resources Association.

Women's Environmental Network (2007) www.wen.org.uk/health/breastcancer.html (accessed 7 August 2007).

World Commission on Environment and Development (1987) *Our Common Future*. Oxford: Oxford University Press.

ZEDFactory (2007) http://www.zedfactory.com/zedfactory_home.htm (accessed 6 August 2007).

4 Valuing the Environment: Environmental Economics and the Limits to Growth Debate

Andrea Revell

Learning outcomes

Knowledge and understanding:

○ of how economic mechanisms affect the environment, both in terms of causing environmental damage, and in attempting to mitigate it

Critical awareness and evaluation:

○ to be able to distinguish between environmental and ecological economics, and the likely impact each has on the environment

○ of the relationship between politics and the economy, in that economic analyses reflect political requirements

---

Introduction

Natural resources are used to produce things that benefit people, but unfortunately these benefits often come at a great cost to the environment. It is important to understand the trade-offs between these benefits and costs, for if continued economic growth means riches for some but dirtier air or fewer forests for all, is it really worth it? Is more economic growth really raising standards of living? Or is there a point where it degrades the quality of life? The fundamental question we thus have to consider is what is the optimal scale of the economy? How big can our economy get before we start to approach the limits to which the earth can support us?

This chapter aims to explore these questions by comparing the views of different schools of economic thought on the relationship between the economy and the environment. The following sections explore how new branches of economics, including the influential school of 'environmental economics', and the more radical approach of 'ecological economics', challenge traditional economic ideas that ignore the environmental impacts of the economy. The chapter concludes by exploring how these new models have influenced environmental policy debates across the globe.

Challenging neo-classical economics

Neo-classical economics has been the dominant paradigm in the economics profession since the late 19th century. This model sees the economy solely in terms of a circular flow of goods and services between producers and consumers in a closed loop. No reference is made to resource depletion or waste as the earth is assumed to have an unlimited capacity to support its population. Kenneth Boulding (1966) famously referred to this as the 'cowboy economy', so called because the cowboys of the North American plains lived on a linear flow of inputs to outputs, from sources to sinks, taking what they wanted from the earth and throwing away the rest. In the cowboy economy, there was no need to recycle anything because resources were assumed to be so abundant.

These assumptions were seriously challenged in the 1960s and 1970s with the advent of the environmental revolution. Seminal books such as Rachel Carson's *Silent Spring* (1962), Paul Ehrlich's *The Population Bomb* (1968) and the Club of Rome's *The Limits to Growth* (1972) foresaw gloomy prospects for the world, due to massive population growth, resource depletion and pollution. The Club of Rome painted a particularly apocalyptic future for the earth if the environmental impacts of the economy were not addressed. The Club of Rome was made up of an informal, international group of scientists, humanists, economists and industrialists who shared a deep concern for the rapid rise in global population, global consumption and industrial output. The stage was being reached, they argued, when this explosion of economic and population growth would exceed the carrying capacity of the earth. Using complex computer modeling programs they predicted that if growth continued at the same rate, the earth would strike its limits by the end of the 20th century.

The book was highly contentious. At that time, there was a general belief in the West that society had been largely released from the constraints of nature, as a result of the massive technological and economic progress that had occurred since the Industrial Revolution, and particularly during the post-war reconstruction period of the 1950s and 1960s. There was a sense that human ingenuity could fix any problem, and that there was a standard Western formula for economic growth and prosperity that could be applied throughout the world. Those proposing the idea of 'limits to growth' were dismissed as alarmist pessimists, enemies of progress and modernity.

However, as environmental problems gradually worsened, the notion of 'limits to growth' began to be taken more seriously. Increased media coverage of environmental issues resulted in growing public awareness of the problems unleashed by industrialisation, and membership of environmental pressure groups soared. Responding to public concern, many governments set up environmental departments and introduced environmental laws to regulate the excesses of industry. The first great world environmental conference, the 'United Nations Conference on the Human Environment', occurred in 1972 in Stockholm, followed by the establishment of the United Nations Environment Programme (UNEP). The 1970s thus heralded what can be considered the 'dawning of environmentalism', when people for the first time began to realise that the environment was not a free good that could be exploited endlessly. This has already been discussed in some detail in Chapter 3.

During the 1970s, environmental non-governmental organizations (ENGOs) were generally sympathetic to the idea that there were limits to growth. They hence allied themselves with radical new branches of economics such as the 'zero-growth' and 'steady-state' schools (see Boulding 1966; Daly, 1973), which supported the idea that the economy was fast approaching the carrying capacity of the Earth, and that production and consumption should therefore be contracted. This meant zero economic growth to avoid breaching ecological limits. While these ideologies were highly influential within the Green movement, they essentially went against the grain of capitalism and free-market economics, and so were wholly untenable to political and business leaders around the world. Developing countries stressed the need to grow their economies to cope with massive population growth and high rates of poverty, while developed countries viewed the idea of contracting production and consumption as political and economic suicide. In supporting the idea of limits to growth, the Green movement hence found itself in direct opposition to a global political economy underpinned by the ideas of traditional neo-classical economics, which unquestioningly viewed growth as 'good'.

In the neo-classical view, growth is the ultimate goal because it assumes that the faster the economy grows, the more material wealth is created, which increases social welfare and results in the utilitarian notion of the 'greatest happiness for the greatest number'. This reflects Adam Smith's invisible hand theorem, which states that the economy is characterised by profit-maximising producers and welfare-maximizing consumers acting in their own self-interest. If all agents act in a self-interested, rational manner, the market's invisible hand will deliver a socially desirable outcome. This outcome is the efficient allocation of resources, which results in the greatest happiness for the greatest number.

Neo-classical economics thus emphasizes the central role of the market in maximizing the social good by allocating all resources – including environmental ones – efficiently. The market is viewed as self-correcting; if resources are exploited too rapidly, they become scarce, prices go up and so usage goes down. Once resources become scarce, producers are encouraged to find substitutes, and this in turn encourages technological innovations that can reduce environmental impacts. The underlying assumption is that the market harnesses individual self-interest for the advantage of society and the environment.

The tension between neo-classical economics and those advocating limits to growth is well illustrated by the debate already illustrated in Chapter 2, between 'cornucopian' economist Julian Simon (1981) and Paul Ehrlich, the celebrated author of *The Population Bomb* (1968). You will remember that Simon won the bet that the price of a basket of metals (chosen by Ehrlich) would go down (indicating plenty) rather than up (indicating scarcity) despite major increases in population growth. Simon thus led credence to his argument that short-term scarcity could be overcome as higher prices motivated inventors and entrepreneurs to search for alternatives.

Mainstream economists continue to follow traditional ideas regarding economy–environment relationships. However, challenges have come from more progressive thinkers in economics who have been concerned about the growing environmental impact of the economy, and have branched off to form their own schools of thought.

Environmental economics

One school that has had a great impact on the environmental policy agenda worldwide is that of 'environmental economics', which emerged in the late 1970s as a challenge to the zero-growth and steady-state ideologies dominant at the time. Instead of zero growth, environmental economics introduced the more complex objective of 'sustainable growth', a concept that was considerably more palatable to those within the halls of power. This perspective argued that the economy could continue to grow as long as Boulding's (1966) 'cowboy economy' was transformed into what Jacobs (1997) terms a 'circular economy'.

Sustainable growth and the circular economy

The idea of a circular economy is founded on the first law of thermodynamics, which states that 'matter cannot be created or destroyed, it can only be transformed'. Under this law, it becomes clear that if matter and energy cannot be destroyed, all inputs must inevitably end up as waste. Material goods do not just disappear once we are finished with them; they are transformed into refuse and pollution. The goal of environmental economics is to minimize this waste by increasing the circularity of the economy, for if wastes can be returned to the economy (i.e. recycled), both resource depletion and pollution can be minimized. The key emphasis in environmental economics is thus to increase the 'eco-efficiency' of the economy by limiting the 'throughput' of production. Advocates of this approach argue that sustainable growth is then possible because the economy has a much smaller impact on the environment.

A major focus of environmental economics is the design of policies which increase resource efficiencies at least cost to the economy. This involves three aspects:

1 commodifying the environment
2 defining the 'optimal' level of environmental protection via cost–benefit analysis
3 creating policies to enable the market to achieve the optimal level of protection.

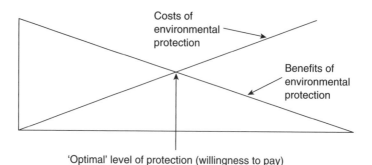

Figure 4.1   The relationship between costs and benefits

Commodifying the environment

Jacobs (1994) highlights that at the heart of environmental economics lies the aim of turning the environment into a commodity. This is because commodification enables environmental goods and services to be incorporated into economic models and analysed like any other market commodity. The process of commodification involves breaking the environment down into its constituent goods and services. For instance, the tropical rainforest provides a range of goods (timber, medicines, food, etc.), and services (climate stabilisation, watershed protection, etc.). Each one of these 'commodities' is then priced according to supply and demand, like any other product or service. In this way, the environment is brought into the market economy.

Finding the optimal level of environmental protection

Environmental economists argue that because environmental protection can be beneficial to society but costly to the economy, benefits and costs should be weighed up to decide the 'optimal' level of protection. This is done in a process called 'cost–benefit analysis' (CBA), which is widely used by policymakers to help them decide which policies to adopt. CBA was developed over the last 30 years by economists as a mechanical way of making decisions about resource allocation. The level of environmental protection that is considered optimal in CBA is the intersection of the costs and benefits curves, as this is where the benefits of environmental protection are maximised at least cost to the economy.

In Figure 4.1, the costs and benefits curves reflect supply and demand curves. As the costs associated with supplying environmental protection go up, demand goes down as people are less and less willing to pay for higher levels of protection.

Environmental valuation methods are essential to cost–benefit analysis, as economists have no other way of comparing benefits with costs except by using financial estimations. In the market, the price of a product or service represents its value or benefit. However, there are no markets – and hence no prices – for many environmental goods and services, so in these instances economists must estimate their value by calculating what people would be willing to pay for that environmental good, or what they would be willing to accept as compensation for environmental losses.

Economists use two methods to assess the value that people ascribe to the environment. The first way is to make an assessment of what is termed 'revealed preferences'. In the case of valuing a national park for instance, economists might include the amount that people pay to travel to a national park, how much they pay to use the park, or how much house prices have gone up around the park. The other method is by 'expressed preferences' using 'contingent valuation'. Contingent valuation involves surveying people on what they are willing to pay for an environmental good or service, or what they are willing to accept as compensation for the loss of that environmental good or service. A monetary value on nature is thus found by researching the expressed preferences of consumers (which of course is constrained by ability to pay – i.e. incomes).

Once a monetary value for environmental commodities has been calculated, the benefits of environmental protection (based on revealed or expressed preferences) may be compared with the costs of that environmental protection, including the opportunity costs of not being able to allocate resources elsewhere. The optimal level of protection is the point where benefits are maximized at least cost.

Creating policies

Once the optimal level of environmental protection has been established using CBA, the next stage is to actually create policies that will achieve that level of protection. Regulations (such as water, land and air quality standards) have traditionally been the central mechanism for protecting the environment. However, since the 1980s deregulation trends have helped to fuel the search for more flexible and cost-effective policy tools. Regulation is often criticized for involving excessive 'red tape' (which can restrict business and discourage innovation), and for requiring a significant amount of funding for enforcement. A central aim of environmental economics has thus been to design more efficient market-based policy instruments, which give producers greater flexibility in deciding how to respond to environmental problems, at the same time as lowering the cost of enforcement. The key market-based instruments (MBIs) used by policymakers today are environmental taxes and tradeable permits.

Environmental taxes   Environmental taxes change the prices of existing goods and services in the market in order to internalise 'environmental externalities'. The idea of an 'environmental externality' relates back to traditional economic ideas. Because the economy is defined solely in terms of production and consumption, with no reference to the ecological systems that underpin it, the environment is considered 'external' to the economy. In setting the price of goods and services, producers thus internalise costs such as capital and labour, but ignore environmental costs such as pollution and waste. In effect, this means that prices in the free-market economy are set artificially low, because they do not reflect the so called 'true social cost of production'. Low prices increase demand, so the response of producers is to increase supply – with a devastating effect on the environment.

Environmental economists highlight that it is because most environmental costs are not internalized within prices that we are facing such acute environmental problems today. Environmental externalities represent a market failure, for they encourage producers and consumers to be wasteful and polluting. For instance, factories in China pump pollution into the air and waterways because they have not had to pay for the impacts of their activity on the environment. Shoppers in Tokyo or New York eat their lunch out of throw-away plastic containers because they can put the packaging in the rubbish to be picked up for free by the municipal authorities. However, although the factory or the shopper does not pay, there is a cost to society. We all ultimately pay if the air is polluted, our waters are toxic and our landfill sites full. Sometimes taxpayers are forced to pick up the tab by funding government spending on pollution control, or they may pay indirectly, for instance by funding treatment for asthmatic patients through their health insurance premiums. But inevitably, as the environment declines, it is the poorest and most vulnerable, not to mention future generations, who end up paying the highest price.

Environmental economics advocates addressing this market failure by making polluters pay directly. This is known as the 'polluter pays principle'. Policies include taxing 'environmental bads' and/or subsiding 'environmental goods', so that polluters have a direct incentive to switch to less environmentally damaging activities. Environmental taxes are thus a way of internalising environmental externalities to ensure that the true social costs of production are revealed in prices. Examples of environmental taxes include carbon taxes (where industry is taxed for every unit of carbon dioxide emitted, which incentivises firms to find energy-efficient and low-carbon alternatives), congestion charging (where motorists pay a fee for entering the 'congestion zone', which encourages them to make fewer journeys in these areas), and fuel duties (where petrol is taxed to encourage motorists to buy more fuel-efficient cars or use their cars less frequently). One example of how taxation could be applied to an environmentally damaging activity – air transport – is given in Box 4.1.

Box 4.1   Taxing air travel

The UK government recently endorsed a massive expansion of air travel capacity in its 'Aviation White Paper' (DTI, 2003), as aviation passenger numbers are predicted to double by 2020. The increased demand for air travel in the UK has been partly driven by the emergence of competitive, low-cost airlines which have put a downward pressure on fare prices and made air travel all the more affordable to the masses. Environmental groups such as Greenpeace, Friends of the Earth and the Green think-tank SERA (Socialist Environment and Resource Association) have criticized government plans to meet this increased demand by building new airports and runways. They argue that these developments are likely to stimulate demand even further, which will have a devastating impact on the environment. Aviation emissions contribute three times more carbon dioxide than any other transport source and are thus one of the major causes of human-induced climate change. The government itself acknowledges that carbon dioxide emissions from aviation could be 25% of Britain's total contribution to global warming by 2030 (DTI, 2003).

Instead of increasing capacity, environmentalists have called for a demand-management approach using environmental taxation to raise the cost of flying. Sewell (2003) highlights that aviation receives £9 billion in hidden subsidies annually, due to the fact that there is no value-added tax on the purchase of aviation fuel, passenger tickets or aircrafts, not to mention aircraft servicing, air traffic control, airline meals or baggage handling. Even the cost of landing slots is subsidized by duty-free airport sales. Because of this, the price of flying is still relatively low and consequently demand for air travel continues to rise at an alarming pace. While car drivers are made to pay directly for the pollution they cause via fuel taxes, vehicle excise duty, congestion charging and road tolls, air passengers are currently exempt from such taxes. Aviation is thus a classic example of an industry which has externalized the environmental and social costs it imposes on society.

Environmental taxation is a favoured policy tool in many countries (particularly Europe), because it is both an efficient and flexible way of engendering environmental reform. Environmental taxes are efficient because the polluter is made to pay directly, and flexible because the polluter can decide whether to pay the tax or find more environmentally friendly alternatives. However, a key problem with environmental taxes is that many polluters may feel that it is easier to pay the tax than to change their behaviour. For instance, despite fuel duties, many people still prefer to use their cars rather than take public transport. Taxes therefore have to be set high to stimulate a behaviour change. Nevertheless, raising taxes is a very contentious political issue, and many politicians fear a backlash from voters. A good example of this is the UK fuel duty protests of 2000, in which the road lobby blockaded motorways to protest at fuel tax rises. The government relented by agreeing not to raise the fuel duty any further.

Environmental taxes can be unpopular with voters for a number of reasons. Firstly, they are seen to believed to penalise the poor more than the rich, as the rich are in a better position to absorb extra costs. They can also be deeply unpopular if it is felt that the taxes have been introduced primarily to increase government coffers rather than to protect the environment. For this reason, there is increasing pressure on governments to 'hypothecate' environmental taxes, which means that the money raised is funnelled directly into services that improve the environment. For instance, in Sweden a tax on fertilisers and pesticides has been spent on research into environmentally sensitive agriculture (ECOTEC, 2001). In the USA, a gasoline tax has been dedicated to improving transportation infrastructure (Parry, 2002). In Australia and the UK, most landfill levies are used to fund waste-management programmes, including recycling (BDA Group et al., 2003). One of the reasons that the UK road lobby was so successful at provoking a public outcry about the rise in fuel duty was that the government had not hypothecated the money to fund improvements in public transport.

Tradeable permits Originally developed in the USA, tradeable permits are another increasingly popular market-based instrument. Tradeable permits work by establishing a target level of environmental quality, defined by 'total allowable emissions'. Permits are then allocated to polluters, with each permit enabling the recipient to emit a specified amount of pollution. Because polluters are allowed to trade these permits among themselves, those who pollute least are rewarded (as they can sell their permits at a profit), and those who pollute most are penalized (as they have to buy extra permits to match their emissions).

Like environmental taxation, the key advantage of tradeable permits is their flexibility, for by letting polluters decide whether to buy more permits or invest in greener alternatives to reduce their emissions, they help to reduce the negative impact of environmental protection on the competitiveness of industry. Tradeable permits are therefore an increasingly popular policy tool within both political and industry circles.

The USA pioneered emissions trading when it established the Emissions Trading Program (ETP) under the Clean Air Act of 1976. Emission permits have since been used in the USA for controlling water and air pollution including lead, CFC, sulphur dioxide and nitrous oxide emissions. In 2005, Europe commenced its own fledgling Emissions Trading Scheme (EETS) which enables member states to buy and sell greenhouse gas emissions permits to help them reach their Kyoto Protocol targets. Box 4.2 explains this programme in more detail.

While tradeable permits are slowly becoming more common in environmental policymaking, they have been criticised by some in the Green movement who believe that giving polluters a permit to pollute is tantamount to giving them the 'right' to pollute, and takes attention away from the need to reduce resource consumption among polluters who can afford to buy more permits.

Box 4.2   Mitigating climate change with carbon trading

Proposals for carbon trading schemes emerged in Europe as a response to the Kyoto Protocol, a 1997 international treaty to mitigate climate change that came into effect in 2005. The nations who have ratified the Kyoto Protocol treaty are legally bound to reduce emissions of six greenhouse gases (collectively) by an average of 5.2% below their 1990 levels by the period 2008–2012, as Chapter 8 on climate change discusses in more detail. The European Emissions Trading Scheme (EETS), which began in 2005, is mandatory for all member states, as a way of meeting their Kyoto obligations to reduce carbon dioxide emissions. It works by capping the amount of carbon dioxide that each country can emit, and then allowing permits to be issued that grant power plants and other large point sources the right to emit a stated amount of carbon dioxide over a fixed time period. This is also known as a 'cap and trade' scheme. Each member state is obliged to submit a National Allocation Plan (NAP), which estimates appropriate caps on emissions for different industries. Once the European Commission has approved the NAP, caps are set in place, and permits are then issued and traded to achieve those caps.

ENGOs became highly critical of the first phase of the EETS (2005–2007) when it became clear that many countries had abused the system, by giving their industries such generous caps that they did not need to reduce emissions beyond current levels.

Environmental groups have thus pressed for much stricter caps in the second phase (2008–2012). The second phase will involve caps for all greenhouse gas emissions, not just carbon dioxide. The European Commission is also considering including aviation as a target industry under the scheme, due to the significant and rapidly increasing emissions produced by this sector.

Aubrey Meyer (2006) has called for a 'contraction and convergence' model for capping global emissions, which places emphasis on the distributional equity of emissions over time. The idea is to ensure that developed countries contract their emissions to such an extent that developing countries can still increase theirs (to keep pace with growing populations), while total global emissions are reduced. Meyer describes it as:

'A straightforward model for an international agreement on greenhouse gas emissions. It sets a safe and stable target for concentrations of greenhouse gases in the atmosphere, and a date by which those concentrations should be achieved, based on the best scientific evidence. The atmosphere being a "global good", C&C declares that all citizens of the Earth have an equal right in principle to emit, and will actually be given an equal right by this future date, the individual allowance for each citizen being derived from the "safe" global target. So from the grossly inequitable situation we have now, per capita emissions from each country will "converge" at a far more equitable level in the future; while the global total of emissions will "contract".' (Meyer, 2006)

Ecological modernization

Since the 1990s, environmental economics has become increasingly influential in defining the response of governments to the environmental crisis. In fact, it has become so influential that many of its core principles have been incorporated into a political programme called 'ecological modernization'. Ecological modernization (EM) is now the policy strategy and discourse adopted by most governments, businesses and reformist environmentalists on the environment (Berger et al., 2001; Revell, 2005). For instance, EM has been described as the new name for environmental politics in the European Union (Pepper, 1999), and is the perspective guiding the World Commission for Environment and Development (WCED) (Strandbakken and Stø, 2003).

Informed by environmental economics, ecological modernization revolves around the imperatives of competitiveness, eco-efficiency, and technological innovation. EM policy strategies also emphasize market-based solutions to environmental problems. A core tenet of this model is that industrialized countries are becoming more environmentally sustainable as a result of technical progress in resource and energy efficiency, which has enabled economic growth to be 'decoupled' from environmental degradation. This idea relates to a concept known in economics as the 'Environmental Kuznets Curve' (or EKC), which is informed by empirical research, indicating an inverted U-shaped relationship between pollution and income per capita, as Figure 4.2 illustrates.

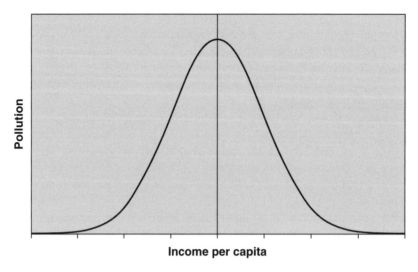

Figure 4.2   The Environmental Kuznets Curve

The curve suggests that while pollution initially increases with rising incomes, it then begins to decrease once a certain threshold of income has been crossed. The central inference is that the wealthier people become, the more their environmental impacts are reduced over the long term as society becomes rich enough to clean up the environmental fall-out of its own growth, and to invest in cleaner production processes. In this model, economic growth is thus part of the solution rather than the problem.

Like environmental economics, advocates of ecological modernization propose that decoupling economic growth from environmental degradation can be achieved only by reducing the material and energy throughput of the economy. Science and technology are seen as crucial institutions for attaining this goal, and policymakers following this path call for major technological innovations to slash the resource use of industry by fourfold and even tenfold. The 'business case for sustainability' – the idea that environmental reform can be profitable, or that 'pollution prevention pays' – is a key rhetorical device used by politicians to encourage technological innovation and a shift towards 'clean production'. Eco-efficiency is emphasized as a way of lowering production costs and increasing profits, while new markets for environmentally friendly products and services are highlighted as potential sources of future growth (Revell, 2007).

Policymakers advocating ecological modernization see the market as the best medium for solving environmental problems, for if market failures are corrected, the most efficient (or eco-efficient) allocation of resources will result. Policy prescriptions in the north thus focus on internalizing environmental externalities via market-based instruments, such as environmental taxes and tradeable permits, and in the south on economic reform and the abandonment of protectionist policies that distort markets.

A groundbreaking report on the economics of climate change, written by the economist Nicholas Stern (2006) and commissioned by the then UK's finance minister Gordon Brown, has caused shock waves around the world by suggesting that climate change is not only the 'greatest and widest-ranging market failure ever seen', but that 'tackling climate change is the pro-growth strategy for the longer term' (Stern, 2006: 1–2). Reflecting the central EM idea that environmental protection is a prerequisite for long-term economic growth, Stern argues that 'mitigation – taking strong action to reduce emissions – must be viewed as an investment ... if these investments are made wisely, the costs will be manageable and there will be a wide range of opportunities for growth and development along the way. For this to work well, policy must promote sound market signals, overcome market failures and have equity and risk mitigation at its core'. He asserts that 'tackling climate change ... can be done in a way that does not cap the aspirations for growth of rich or poor countries' (Stern, 2006: 1–2).

Despite EM's growing prominence within mainstream environmental politics, the principles which constitute its foundation have been challenged by other emerging and more radical branches of economics which have called for a less reductionist and more precautionary approach to environmental problems. The remainder of this chapter explores what one particular branch – 'ecological economics' – offers as an alternative and competing paradigm to that presented by environmental economists.

Ecological economics

> Ecological economics is a transdisciplinary effort to link the natural and social sciences broadly, and especially ecology and economics ... One of the basic organizing principles of ecological economics is ... a focus on (the) complex interrelationship between ecological sustainability (including system carrying capacity and resilience), social sustainability (including distribution of wealth and rights, social capital and coevolutionary preferences), and economic sustainability (including allocative efficiency in the presence of highly incomplete and imperfect markets). (Costanza, 2003: 1)

Ecological economics is in part an outcome of recent developments in systems ecology, which focuses on the complex non-linearity of ecological processes and the resilience of ecosystems to maintain themselves in the face of stress and shock (Turner et al., 1995). Ecological economics thus has its intellectual roots in the natural sciences, which is a key reason why some mainstream economists refuse to regard it as a legitimate branch of economics. While the neo-classical purist implicitly views the environment as external to the economy, ecological economists see the economy as contained within the ecological and biospheric systems of the earth, and therefore economics is viewed as a subset of ecology. Key authors in ecological economics include prominent economists such as

Herman Daly (1995) and David Pearce (1999), as well as ecologists such as Robert Costanza (1991) and Anne-Marie Jansson (1984).

To address the differences between ecological and environmental economics, it is useful to explore the challenges that the former poses for the latter. Key fault-lines between the two branches include:

1   the degree to which a mechanistic, reductionist versus a holistic, coevolution-ary approach is taken
2   the sustainability of a circular economy
3   safe minimum standards and the substitutability of human and natural capital
4   quantitative growth versus qualitative development
5   the valuation of environmental resources.

Reductionism versus coevolution

In contrast to the reductionist approach taken by environmental economics, eco-logical economics sees all relationships (economic, social, ecological) as endoge-nous and coevolutionary. Norgaard (1997) highlights that natural and human systems cannot be separated from one another; in fact, they are so intimately interconnected that they evolve together and reflect one another. In the ongoing feedback between evolving systems, a change in any one system affects the other, and the characteristics of each system places selective pressure on the rest. Ecological economics thus emphasizes the need to understand the integral dynamics between social, economic and ecological systems, and humankind's role in shaping the biological as well as cultural evolution of the planet. By treat-ing the environment as separate from humankind, environmental economics is accused of undermining the dynamic and unified relationship between humans and nature.

Ecological economists also regard it as folly to assume that we can leave deci-sions about resource allocations to the market and to 'benign' interventions by government (using environmental valuation and CBA) to manage the environ-ment on behalf of society. Norgaard (1997) highlights the inherent unpredictabil-ity of coevolutionary processes, as nothing can be said with certainty to determine anything. Power is diffuse in the coevolutionary model – all can and will transform in unpredictable ways, and human agency is just one aspect among many others that determines future development. Challenging the idea that the environment can be 'managed', or that environmental change can be predicted with any accuracy, ecological economists advocate a far more ecocentric and precautionary approach to economic analysis and policymaking. Conservation and environmental protection measures – particularly protection of ecosystem resilience – take precedence, unless the cost to society is unacceptably large (Turner et al., 1995).

The sustainability of a circular economy

One aspect of this ecocentrism is that while environmental economists do not believe in predetermined limits to growth, ecological economists assert that there are environmental limits that should constrain economic activity even in a circular economy. Using Georgescu-Roegen's (1971) 'entropy'-based model, which forms a core intellectual foundation for ecological economics, Daly (1991, 1992) argues that there are limits to physical growth even in the most resource-efficient economy, because perfect recycling is precluded on thermodynamic grounds. Daly highlights that environmental economists have rather missed the point with the circular economy, for while they may have grasped the implications of the first law of thermodynamics and the limits it imposes, they have not grasped the implications of the second law. The second law of thermodynamics states that 'when energy is transformed from one state to another, some energy is lost as heat'.

This law highlights that the transformation of energy or matter involves some degradation of quality: a shift from order to disorder (or entropy). The circular economy assumes that everything goes round in a closed loop with nothing wasted. However, the law of entropy tells us that in reality there will always be some waste. Economic activity involves transforming materials from one state to another, and this dissipates energy in the process. Economic activity thus necessarily increases entropy (disorder) in the universe.

Under the second law of thermodynamics, every day the universe gets more and more disordered. Any local increase in order (e.g. production of a good) is paid for by an increase in disorder elsewhere (e.g. by-products of toxic waste, pollution). Any system without an external energy source increases entropy. There is thus a major flaw in the circular economy; a closed system will eventually use up all its available energy and turn it into waste. For example, recycling requires energy to power treatment plants. As long as the recycling process is powered by fossil fuels, it will contribute to air pollution. Recycling therefore reduces one environmental problem (waste) only at the cost of increasing another (air pollution). Because of the entropy problem, ecological economists like Daly (1992) argue that the circular economy is simply not sustainable, and that solving the entropy problem requires the creation of a renewable energy economy. By harnessing solar power (the earth's external energy source), energy can be added into the system, replacing what has been lost. Using solar power, the circular economy then becomes a feasible goal, as the energy required to recycle is unlimited and environmentally benign. The real significance of renewable energy is thus not that it does not run out but that it does not generate waste (Jacobs, 1997).

Safe minimum standards and the substitutability of human and natural capital

Ecological economics emphasizes the extreme scientific uncertainty surrounding complex ecological processes and functions, arguing that because we do not know enough about the long-term impacts of pollution and other environmental hazards, and because new effects can occur due to the inherent unpredictability and interrelatedness of environmental systems, it may not be possible to identify sustainable levels of economic growth, or the 'optimal' level of environmental protection as emphasized in environmental economics. As ecosystems do not always clearly signal when their carrying capacity or assimilative capacity has been breached, market prices cannot be relied upon to indicate whether a system is approaching its threshold of resilience. Rather than a simple cost–benefit approach, ecological economists thus advocate the application of the 'precautionary principle' in environmental policymaking, which involves the adoption of 'safe minimum standards'.

Safe minimum standards, such as quotas or limits on resource use, emission permits, ambient standards or hunting seasons, are a precautionary instrument to safeguard 'critical natural capital' – those environmental goods and services that are critical to the healthy functioning of ecosystems and biodiversity. By preserving a critical minimum stock of natural capital, safe minimum standards are seen as a buffer against unexpected environmental behaviour and the risk of irreversibility.

In contrast, environmental economics adopts less stringent ideas of environmental conservation, as no aspect of the environment is considered inviolable – including critical natural capital. Instead, environmental economics takes the view that depletion of natural capital is justified as long as the 'Hartwick rule'[1] is observed, i.e. that there is an equivalent increase in human capital which ensures that the total capital stock remains the same. This means that the income derived from natural resources must be invested in other forms of capital which can yield the same amount of income. Thus, as long as fossil fuel or tropical rainforest exploitation results in a corresponding increase in human capital such as roads, buildings, electricity, schools or incomes, the exploitation is deemed acceptable to environmental economists. This is known as a 'weak' perspective on sustainability.

Reflecting Julian Simon's cornucopian beliefs, those advocating weak sustainability argue that, as natural capital becomes more and more scarce, there are increased incentives to search out alternative supplies or artificial and natural substitutes. For example, dwindling stocks of hardwood timber have led to softwood and manufactured substitutes. Environmental economists are therefore optimistic about the possibility of finding suitable technical substitutes for natural capital. In contrast, ecological economists take a strong sustainability approach, which assumes that technology will not necessarily be able to remove resource constraints, and that there are no substitutes for some (critical) natural capital. Instead of viewing natural capital as interchangeable with human capital, ecological economics views human capital as derived from natural capital and therefore dependent on natural capital. The ecological health of the planet therefore takes precedence over the economic needs of humans.

Quantitative growth versus qualitative development

Daly (1987) argues that environmental economics confuses quantitative economic growth (i.e. increases in the biophysical scale of the economy, as measured by national accounts or Gross Domestic Product (GDP)), with qualitative development in the non-physical characteristics of the economy (indicated by an improvement in the 'quality of life'). Daly and other ecological economists highlight that the neo-classical model of GDP measures a nation's prosperity by the volume of goods and services bought and sold, but ignores the social and environmental cost of doing business. They argue that the use of GDP as a measure of the standard of living is misconceived if increased consumption leads to degradation of natural capital, which in turn results in a reduction in the quality of life and general well-being of society. For instance, a study by Daly and Cobb (1987) found that economic growth in the US during the 1950s and 1960s caused welfare per capita to increase, but further economic growth in the 1970s and 1980s actually caused overall welfare to decline. They concluded that after a critical threshold, the negative aspects of affluence started to outweigh the positive aspects, and that further economic growth could lead to a reduction in the quality of life.

Marylin Waring's celebrated book, *If Women Counted: A New Feminist Economics* (1988), outlines a scathing critique of GDP for ignoring the environment, domestic and unpaid work, and subsistence production. She highlights how GDP records environmental degradation as ultimately welfare producing rather than reducing, as oil spills and pollution are counted as having a positive effect on economic growth, as Box 4.3 demonstrates with reference to the Exxon Valdez oil spill.

Box 4.3   *Exxon Valdez* oil spill as 'welfare producing'

On 24 March 1989, the supertanker *Exxon-Valdez* tragically crashed into reefs off the port of Valdez on the Alaskan coast, dumping an estimated 11 million gallons of crude oil into the sea in a matter of 7 or 8 hours (Buchholz and Rosenthal, 1998). The oil spill was one of the worst in history, devastating the ecology of Alaska's pristine Prince William Sound. A massive clean-up operation ensued, at a total cost of US $2 billion. Since GDP records every monetary transaction as positive, the *Exxon Valdez* oil spill was by inference counted in the national accounts of the United States as welfare producing rather than welfare reducing.

Because of these criticisms of GDP, ecological economics advocates the use of alternative indexes such as Green Gross Domestic Product (Green GDP) or Green National Product (GNP), which factor in the environmental consequences of economic growth and therefore offer a more realistic measure of the quality of life. Examples include:

- Nordhaus and Tobin's (1972) MEW (Measure of Economic Welfare), which subtracts the costs associated with environmental 'bads' (such as pollution) from GDP, while adding qualitative benefits such as the value of services (versus the sales price of products), the value of leisure time, household production and government spending.
- Daly and Cobb's (1987) Index of Sustainable Economic Welfare (ISEW), which weights personal consumption with factors such as the cost of environmental degradation, depreciation of natural resources, capital adjustments, income inequality and domestic labour.
- Pearce and Atkinson's (1993) 'genuine savings' model, which adjusts GNP by taking account of resource depletion, environmental degradation and changes in environmental services to measure true savings.

The valuation of environmental resources

Ecological economics challenges environmental economics' reliance on contingent valuation and cost–benefit analysis in deciding environmental policy. Because the latter puts a monetary value on environmental goods and services by assessing what people are willing to pay for them, environmental value is ascribed mainly in terms of nature's utility or value to human beings. However, ecological economists emphasize not just the instrumental but the intrinsic value of nature, i.e. its inherent worth regardless of what humans ascribe to it. Ecological economists thus question whether it is possible to put a monetary value on the environment at all. They also highlight that aggregating private willingness to pay may not adequately capture the 'true' social value of healthy ecosystems and interrelationships.

Ecological economists have much sympathy with Mark Sagoff's famous and devastating critique of contingent valuation in his book *The Economy of the Earth* (1988). Sagoff's essential argument is that people value the environment as citizens not consumers. People have economic preferences as consumers, but they also have moral preferences as citizens. Because the environment is a 'public good', people tend to react more as citizens than consumers, and this makes it very difficult for them to put a monetary value on nature. This is evidenced by the problems encountered in contingent valuation surveys. Firstly, large numbers of people refuse to participate in such surveys. Secondly, the meanings of those that do respond are not always clear, as people are unsure about how to reply to questions such as 'how much are you willing to pay for the protection of the blue whale?'. Thirdly, what people are willing to pay for environmental protection is usually less than the compensation they would need for its loss. Sagoff highlights that this is an important finding, as it means that willingness to pay consistently underestimates environmental losses and therefore allows too much environmental degradation. Sagoff concludes that contingent valuation may thus misrepresent the way in which people value the environment.

Confirming Sagoff's analysis, Burgess et al.'s (1999) study of a contingent valuation exercise, regarding the Pevensey Levels in southern England, found that the participants had difficulty in fragmenting their experiences of nature in the way required by the response analysts. Moreover, participants expressed anger and concern when informed that their responses could be used as a basis for environmental policymaking. They concluded that contingent valuation exercises were fundamentally flawed because participants found valuing nature an alien concept.

Another problem with contingent valuation and cost–benefit analysis (CBA) is that current preferences are not necessarily a good indication of true welfare (O'Neil, 1993). For instance, people may prefer shopping malls to wetlands because they have no idea how ecologically important wetlands are. Consequently, what people are willing to pay for wetland preservation may be comparatively little compared to the value they ascribe to having their favourite shops in close proximity. CBA is therefore criticized for undervaluing environmental resources.

Detractors of cost–benefit analysis conclude that the central problem is that it can – and does – result in immense environmental degradation, particularly as the benefits of 'development' are so often considered to outweigh the costs to the environment. This is particularly true in developing countries, whose development needs are greatest.

Ecological economists thus challenge the central emphasis of CBA in environmental policymaking, arguing that environmental economics' reliance on preference-based valuations may result in aesthetically attractive environmental goods getting assigned higher prices than life-supporting critical natural capital. Ecological economics supports the protection of critical natural capital via standards and regulations rather than monetary valuation methods.

As part of its focus on social sustainability, ecological economics also supports the expansion of the field of economics to include more participative methods of environmental decision making than cost–benefit analysis. This may include consensus or democratic decision-making methodologies where stakeholders participate in reaching objectives and priorities about resource allocations that affect them, as in the case of lay panels, round-table meetings or citizens' juries (Aldred and Jacobs, 2000). The narrow focus on CBA by environmental policymakers is criticized as exclusionary and elitist, assigning the ultimate decision-making power to non-elected policymakers and economic 'experts', instead of encouraging wider public involvement in solving environmental problems. Ecological economists thus support the political model of 'deliberative democracy' and the reform of institutions to increase stakeholder participation in environmental decision making, especially those affected by a particular issue (Costanza et al., 1997). Ecological economists view a sustainable society as requiring a strong democracy where different viewpoints, values and aims are aired in free and open debates, and where citizens are well informed and given an equal standing with so-called experts (O'Hara, 1996; Prugh et al., 2000).

Summary

Environmental economics is one of the fastest growing academic fields and, as the dominance of ecological modernization discourse attests, its analysis has already influenced policymakers around the world. This is a significant achievement, given that only a few decades ago environmental considerations held little or no sway within the halls of power.

Environmental economists view sustainable growth as entirely possible, as long as the circularity of the economy is increased which, they argue, will decouple environmental degradation from economic growth. If environmental externalities (such as pollution and waste) are internalised within prices, the market system can effectively deal with environmental problems by limiting demand. With the help of natural resource economics, scarce resources can be used sustainably. Moreover, environmental economists argue that if environmental goods and services are given a monetary value in economic decision making, they are more likely to be protected.

In favouring the notion of 'sustainable growth' and the commodification of the environment, environmental economists are sometimes seen as materialist philistines by those in the Green movement. Environmental economists, on the other hand, see themselves as true environmental champions because putting a monetary value on environmental goods and services ensures that they count in the 'real world' of finance and economic decision making.

As we have explored in this chapter, there are strong criticisms of environmental economics emerging from more radical schools of thought, such as ecological economics. While still on the fringes of academic inquiry, the emergence of this school indicates an increasing recognition among scholars that an understanding of the dynamic and complex physical systems on which economic activity depends is crucial to furthering the field of economics. Some of the critiques emerging from ecological economics are slowly being integrated into environmental economics (such as the need for a renewable energy *and* circular economy). Nevertheless, some would argue that these very different approaches represent fundamentally opposing paradigms and therefore may never be fully reconciled.

Note

1. The Hartwick rule is named after the economist John Hartwick, who defined his theorem in Hartwick (1977).

References

Aldred, J. and Jacobs, M. (2000) 'Citizens and wetlands: evaluating the Ely citizen's jury', *Ecological Economics*, 34: 217–32.

BDA Group and McLennan Magasanik Associates (2003) 'The potential of market-based instruments to better manage Australia's waste streams', www.deh.gov.au/settlements/publications/waste/mbi/study-2003/pubs/study.pdf (accessed 5 December 2006).

Berger, G., Flynn, A. and Hines, F. (2001) 'Ecological modernization as a basis for environmental policy: current environmental discourse and policy and the implications on environmental supply chain management', *Innovation*, 14(1): 55–72.

Boulding, K. (1966) 'The economics of the coming spaceship earth', in H. Jarrett (ed.) *Environmental Quality in a Growing Economy*, Baltimore, MD: Johns Hopkins University Press. p. 7.

Buchholz, R. and Rosenthal, S. (1998) *Business Ethics*. Englewood Cliffs, NJ: Prentice Hall.

Burgess, J., Harrison, C. and Clark, J. (1999) 'Respondents' evaluations of a contingent valuation survey: a case study based on an economic evaluation of the wildlife enhancement scheme, Pevensey Levels in East Sussex', *Area*, 30(1): 19–27.

Carson, R. (1962) *Silent Spring*. Boston: Houghton Mifflin.

Club of Rome (Donella H. Meadows, Dennis L. Meadows, Jorgen Randers, and William W. Behrens III) (1972) *The Limits to Growth*. New York: Universe Books.

Costanza, R. (ed.) (1991) *Ecological Economics: The Science and Management of Sustainability*. New York: Columbia University Press.

Costanza, R. (2003) 'Ecological economics is post-autistic', *Post-Autistic Economics Review*, 20(2): 1–3.

Costanza, R., Cumberland, J., Daly, H., Goodland, R. and Norgaard, R. (1997) *An Introduction to Ecological Economics*. Boca Raton, FL: St Lucie Press.

Daly, H. (1973) *Towards a Steady State Economy*, San Francisco, CA: W.H. Freeman & Co.

Daly, H. (1987) 'The economic growth debate: what some economists have learned but many have not', *Journal of Environmental Economics and Management*, 14: 323–36.

Daly, H. (1991) 'Elements of environmental macroeconomics', in R. Costanza (ed.), *Ecological Economics: The Science and Management of Sustainability*. New York: Columbia University Press. pp. 32–45.

Daly, H. (1992) 'The economic growth debate: what some economists have learned but many have not', in A. Markandya and J. Richardson (eds), *The Earthscan Reader in Environmental Economics*, London: Earthscan, pp. 36–49.

Daly, H. (1995) 'On Nicholas Georgescu-Roegen's contributions to economics: an obituary essay', *Ecological Economics*, 13: 149–54.

Daly, H.E. and Cobb, J.B. (1987) *For the Common Good: Redirecting the Economy towards Community, the Environment, and a Sustainable Future*. Boston, MA: Beacon Press.

Department of Trade and Industry (DTI) (2003) 'The future of air transport', Aviation White Paper, http://www.dft.gov.uk/stellent/groups/dft_aviation/documents/division homepage/029650.hcsp (accessed 31 October 2006).

Ehrlich, P. (1968) *The Population Bomb*. New York: Ballantine Books.

ECOTEC (2001) 'Study on environmental taxes and charges in the EU', ec.europa.eu/environment/enveco/taxation/annex3.pdf (accessed 5 December 2006).

Georgescu-Roegen, N. (1971) *The Entropy Law and the Economic Process*. Cambridge, MA: Harvard University Press.

Hartwick, J.M. (1977) 'Intergenerational equity and the investment of rents from exhaustible resources', *American Economic Review*, 66 (December): 972–4.

Jacobs, M. (1994) 'The limits to neo-classicism: towards an institutional environmental economics', in Michael Redclift and Ted Benton (eds), *Social Theory and the Global Environment*. London and New York: Routledge. pp. 67–91.

Jacobs, M. (1997) T*he Green Economy: Environment, Sustainable Development and the Politics of the Future*. London: Pluto Press. pp. 86–116.

Jansson, A.M. (ed.) (1984) *Integration of Economy and Ecology: An Outlook for the Eighties*. Stockholm: University of Stockholm Press.

Meyer, A. (2006) 'Contraction and convergence: global solutions to climate change', RSA 16th February, http://www.thersa.org/acrobat/meyer_160206.pdf (accessed 9 November 2007).

Myers, A. (2006) 'The fair choice for climate change', http://news.bbc.co.uk/1/hi/sci/tech/4994296.stm (accessed 7 December 2006).

Nordhaus, W.D. and Tobin, T. (1972) 'Is growth obsolete?', *Economic Growth: Research General Series 96F*, National Bureau of Economic Research. New York: Columbia University Press.

Norgaard, R.B. (1997) 'A coevolutionary environmental sociology', in M. Redclift and G. Woodgate (eds), *International Handbook of Environmental Sociology*. Cheltenham, UK: Edward Elgar.

O'Hara, S.U. (1996) 'Discursive ethics in ecosystems valuation and environmental policy', *Ecological Economics*, 16(2): 95–107.

O'Neil, J. (1993) 'Justifying cost–benefit analysis: arguments from welfare', in his *Ecology, Policy and Politics: Human Well-being and the Natural World*. London: Routledge. pp. 62–82.

Parry, I. (2002) 'Is gasoline undertaxed in the United States?', *Resources*, 148: 28–33.

Pearce, D.W. (1999) 'Economic valuation', in D.W. Pearce (ed.), *Economics and the Environment: Essays in Ecological Economics and Sustainable Development.* Cheltenham, UK: Edward Elgar.

Pearce, D.W. and Atkinson, G. (1993) 'Capital theory and the measurement of sustainable development: an indicator of weak sustainability', *Ecological Economics,* 8: 103–8.

Pepper, D. (1999) 'Ecological modernization or the "ideal model" of sustainable development? Questions prompted at Europe's periphery', *Environmental Politics,* 8(4): 1–34.

Prugh, T., Constanza, R. and Daly, H. (2000) *The Local Politics of Global Sustainability.* Washington, D.C. and Covelo, CA: Island Press.

Revell, A. (2005) 'Ecological modernization in the UK: rhetoric or reality?', *European Environment,* 15: 344–61.

Revell, A. (2007) 'The ecological modernization of small firms in the UK's construction industry', *Geoforum.*

Sagoff, M. (1988) *The Economy of the Earth: Philosophy, Law, and the Environment.* New York and Cambridge: Cambridge University Press.

Sewell, B. (2003) 'The hidden cost of flying', Aviation Environment Federation, http://www.airportwatch.org.uk/publications/Hidden%20Cost%20Final.pdf (accessed 31 October 2006)

Simon, J. (1981) *The Ultimate Resource.* Princeton, NJ: Princeton University Press.

Stern, N. (2006) 'Stern review on the economics of climate change', www.hm-treasury.gov.uk/independent_reviews/stern_review_economics_climate_change/stern_review_report.cfm (accessed 11 December 2006).

Strandbakken, P. and Stø, E. (2003) 'Eco-labels and ecological modernization: consumer knowledge and trust in eco-labels in four European countries'. Paper presented at the ETE workshop, Turin, 31 January.

Turner, K., Perrings, C. and Folke, C. (1995) 'Ecological economics: paradigm or perspective'. CSERGE Working Paper GEC 95-17, University of East Anglia, UK. pp. 1–33.

Waring, M. (1988) *If Women Counted: A New Feminist Economics.* San Fransisco, CA: Harper and Row.

5 Geoinformation Technology and the Environment

Adrian Combrinck, Mike Turner and Robert Wright

Learning outcomes

Knowledge and understanding of:
- ○ what geoinformation is
- ○ what Geographical Information Systems (GIS) are and what remote sensing is
- ○ the essential principles of how GIS and remote sensing operate
- ○ the increasing convergence of a wide variety of technologies
- ○ a number of examples worldwide in which geoinformation technology is utilised
- ○ the relationship between technologies and demographic, social and economic processes, particularly in relation to the environment

Critical awareness and evaluation:
- ○ you will have learnt that the use of technology can be problematic
- ○ you will have come to realise the importance of enabling the population at large to interact with technology
- ○ you will appreciate the importance of initiatives to increase public participation in issues which lend themselves to technological solutions
- ○ you will have developed the facility to think critically about the issues raised in the chapter

---

Introduction: What is Geoinformation?

All over the world, governmental, commercial and academic organisations are using, processing or supplying geoinformation for a vast range of applications. Why is it that geoinformation matters so much?

As Longley et al. (2005a) note, *'Almost everything that happens, happens somewhere. Knowing where something happens can be critically important'*.

This chapter aims to explain what is meant by geoinformation and to introduce the technologies that use it in relation to the environment. A careful examination of the apparently unremarkable picture in Figure 5.1 begins this process.

Figure 5.1
Source: Groundwork Thames Valley

Look closely and you will see that the picture is not a photograph. It is a computer-generated picture of a landscape that does not yet exist. Even so, it is derived from a landscape that *does* exist (part of the Grand Union Canal to the west of London), and that was created to show how that landscape will look after planned modifications. The information about the existing landscape has come from a variety of sources which are examined below.

First, the general locational information has come from a digital map (a map that can be manipulated by a computer). Then the relief has been identified by data from one satellite image (see the second part of this chapter, Remote Sensing, p. 98) and the land use from another one. The data for precise, detailed, locations of small features have been identified using information from the Global Positioning System (GPS), which is another satellite system (see p. 80). Detailed information about what is to be found at various locations (things like how deep the water is, what material the towpath is surfaced with) are held in a geographical information system (GIS) (see p. 81). All these sources of data constitute geoinformation. It is geoinformation because they all refer in some way to locations on the Earth's surface and technology is employed to convert the information into the picture.

The realistic-looking vegetation and the reflections in the water are not derived from geoinformation, but are a further example of the use of technology

Figure 5.2   Extract from the SAUL project fly-through
Source: Groundwork Thames Valley

because they have been created using software specialising in representing natural materials.

Another important technological aspect of the picture in Figure 5.2 (unfortunately one not easily demonstrated on a paper page) is that it is a still image from a moving sequence, a 'fly-through', which gives the viewer an impression of flying through the landscape. The user can stop and start the fly-through, alter the 'altitude' and control the 'direction' of flight, as well as determining the level of detail. The light grey line is the boundary of the project area described on p. 96–98. Figure 5.1 is part of a *walk*-through, similar in principle to a fly-through.

Apart from the sources of geoinformation and the technologies, another very important aspect of the fly-through is the reason why this visualisation was created, which was to communicate proposed plans to the public and to encourage them to participate in the final decisions about the proposal. The public participation process is considered in more detail on p. 96. The visualisation is part of the SAUL project, a big environmental improvement project near Heathrow airport, which is described in more detail on p. 96–98. The information is delivered over the Web to allow maximum exposure and, for those without access to a computer, it can be viewed at the project team's offices.

Technological development

We now consider two main technologies: Geographical Information Systems (GIS) and remote sensing. However, although the chapter is organised in this way, the analysis of the pictures in Figures 5.1 and 5.2 has shown that the technological contribution to environmental understanding extends considerably beyond these two methods and the boundaries between all the technologies are becoming increasingly blurred. This section points to the many ways in which technology can contribute to understanding the environment itself and to being an aid for environmental decision makers. Also, it is not just a question of the technology that is important, it is also the perceptions of it and the uses it is put to by society at large – the public, government, business, for example – which affect the outcomes of technological implementation.

At one time, less than 20 years ago, GIS and remote sensing were different packages located on a single computer. It may have been a networked computer but even that was relatively unusual then. Now they are no longer restricted to 'stand-alone' systems. We have already mentioned information about the SAUL project conveyed over the Web, but GIS and remote sensing 'engines' themselves may be delivered through the Web and, more widely, their 'products' are Web-based.

Many people now use GIS and remotely sensed products without being aware of what they are. GIS is now common in local government websites which are frequently interactive. You go to the site to find out, for example, about planning developments, or flood risk, or where the open spaces are. Mapping sites such as www.multimap.com, www.streetmap.co.uk, www.google.com/maps (for the USA) or www.google.co.uk/maps (for the UK) will not only show you where an address or a street is but, if you ask it to, will tell you about nearby businesses and facilities such as hotels, shops or schools. Google Earth (downloadable from Google) and http://local.live.com are remarkable demonstrations of technological advancement in geoinformation processing and visualisation.

Mobile phones can access the internet too, so much of what is discussed above is available on the move. However, at present some geoinformation is still too big for mobiles. Then there are GPS (global positioning systems). This is satellite technology in which a GPS receiver communicates with a constellation of satellites and, by receiving signals from several of them, is able to identify the location of the receiver. GPS have been around for a while now and have been widely used in conjunction with GIS. Now, however, they are entering the mainstream. 'SatNav' is becoming increasingly common in cars but to work, the GPS has to be combined with a GIS to display a map. The GPS giving a driver just the latitude and longitude of the location would not be all that helpful.

Remote sensing is also developing rapidly. Resolution from space (for the public, it is higher for the military) is now around one metre (i.e. objects one metre across can be identified). But remote sensing now increasingly includes digital images, at very high resolution, sensed from aircraft. In much of the geoinformation now being distributed and used, digitally sensed data and GIS are seamlessly integrated.

Criticisms of technology

It has long been recognised that technologies have limitations. At one simple level, there is the problem of user error. If invalid data are input into analysis technology then the output will be invalid too. This has long been known in the computer world as the GIGO principle ('garbage in, garbage out').

There are continuing problems of how technologies represent the real world. The representation cannot be an exact replica of what is there (every blade of grass?). In order to represent the world, either by paper map or digitally, the real world has to be sampled and modelled, and some degree of homogeneity has to be introduced. That is, regions, or areas, have to be identified which are the same throughout and different from neighbouring areas. Both sampling and homogeneity introduce differences from the real world. Also, in creating regions, technologies have to create boundaries, which themselves are not realistic. For example, soil types do not have clearly defined edges; they merge, one into another. Attempts to handle such problems have led to the concept of fuzziness or vagueness with which interpreters of geoinformation have to contend.

All the technologies are used not just to deliver information but as an aid to decision makers. The use of technology in this way has led to criticism, both because of its technical limitations, and because the use made of it in decision making may not be transparent to those affected by it. This latter point is particularly important. Attempts to improve communication by, for example, delivery over the Web may still not be accessible to all. Even now, not everyone has access to a computer and not everyone knows how to use one. These problems, and examples of ways of dealing with them, are examined in greater detail in the GIS section of the chapter.

Finally, criticism of the technology has led to the emergence of 'geoscience' or 'GIScience' (Geographic Information Science), an alternative meaning for the acronym 'GIS'. One way of looking at these developments is to see GISystems as the practical application of the technology and GIScience as the academic study of the principles involved. Many of these issues are particularly well discussed in the first chapter of Longley et al. (2005a) and, from a more philosophical perspective, by Schuurman (2004).

What are Geographical Information Systems (GIS)?

Many textbooks on GIS begin with an attempt to define GIS. However, it is perhaps more effective for people completely new to the subject to begin with some characteristics of GIS. Definitions can then be considered later.

One way to think of GIS is that they have three basic capacities:

1   to store and manage data
2   to represent the data visually, such as by drawing maps
3   to analyse the data and to perform mathematical computations.

However, the most important feature of GIS, which distinguishes them from other types of data-handling programs, is that the information they manipulate is *geo*information.

Another helpful approach to understanding what GIS are is to identify the types of question that GIS can answer. Heywood et al. (2006: 3) suggest the following:

- Where are particular features found?
- What geographical patterns exist?
- Where have changes occurred over a given period?
- Where do certain conditions apply?
- What will the spatial implications be if an organisation takes a certain action?

In this list, the key words are: 'where', 'geographical', and 'spatial'.

Databases and database management

GIS store data in a database. For those unfamiliar with databases, Box 5.1 outlines the basic principles of databases and database management.

Box 5.1   Databases and database management systems

A database is simply a collection of data. On its own, it does not have any meaning. In order to be of use, the data needs to be 'managed'. This management often involves organising the data in such a way that it can be readily understood and, particularly, so that it can be interrogated or queried, in order to extract or retrieve desired subsets of the data. Such management systems are known, not surprisingly, as database management systems, or DbMS for short. Confusingly, however, these systems are often loosely known simply as 'databases', which, though not strictly accurate, is very common. Thus, Microsoft Access, for example, though a DbMS, is usually known as a 'database' or 'database package'.

A simple example of database management is an address book. If you have a notebook with names and addresses in it but randomly scattered throughout the book, it is not much use if you want to find a particular address. If, however, you have an address book in alphabetical order of pages, then retrieval becomes possible. However, there are limitations. What happens when a particular letter page is full? Also, what if you want to know the phone numbers of all the people in your book who live in a particular place? Computer-based DbMS solve these (and many more) problems because they are able to scan through huge volumes of data in a very short time. This means that data can be entered in a way which enables many combinations of information to be retrieved.

Usually, though by no means always, a computer database is viewed as a table with rows and columns, and is therefore often known as a tabular database.

Georeferencing

The type of DbMS described in Box 5.1, of which there are many in general use, and which can be extremely complex and sophisticated, does not have information about the locations of the data. Such DbMS may appear to contain locations, such as an address, but without a map for reference, the address itself says nothing about where it is on the Earth's surface. The vital characteristic of GIS databases is that they contain information about precise locations relating to the data stored. This aspect of GIS databases is known as 'georeferencing'. It is georeferencing which, among other things, allows GIS to draw maps. However, georeferencing on its own is not enough. GIS databases need two kinds of information: one is the georeferencing data and the other is about what is to be found there.

Thus, a simple way of conceptualising GIS databases is as two parallel-linked databases, one containing the georeferencing (the 'spatial' database) and one simple tabular database as described in Box 5.1, containing other information, usually known as attributes (but see raster data models below, for alternative methods).

It is important to remember that a GIS stores *data*, not maps. The user can ask the GIS to display maps but maps are not necessarily required for all GIS operations.

Data models

There are several ways in which GIS represent spatial data. However, only the three principal ones will be described here. Of the three, the two main ways in which GIS handle georeferenced data are 'vector' and 'raster' data models. The two, described below, work in different but complementary ways, and both are necessary for different GIS functions.

The spatial databases, in themselves, say nothing about what is present at the locations identified by the models. The GIS, therefore, must contain information about what is to be found at these locations. This data, referred to as 'attribute data', is handled in different ways by vector and raster-based GIS.

Vector data models

The spatial database contains data about specific locations in the form of coordinate pairs. At the most fundamental level, the coordinates refer to latitude and longitude. However, surrogates for latitude and longitude, such as national grids, are often used. The coordinates identify intersections of lines.

A single coordinate pair identifies a *point* and a succession of such pairs identifies a *line* (i.e. a succession of points). If the interval between the points is short enough and the points numerous enough, the line appears to be a smooth curve. If the point at the end of a line has the same coordinates as the first one, then it identifies an *area* (because the end of the line joins up with the start). Areas are usually referred to as 'polygons' because they consist of a large number of short lines, as described above. A careful examination of Figure 5.3 will help clarify this explanation. If you are unfamiliar with coordinates notice, in the first diagram, how the system works. For point number 1, you will see that the first number of the coordinate pair, 01, is one of the numbers along the bottom of the grid, while the second number, 03, is one of the numbers up the side. You will see that this identification of coordinates operates through all three diagrams.

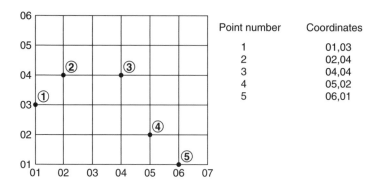

Point number	Coordinates
1	01,03
2	02,04
3	04,04
4	05,02
5	06,01

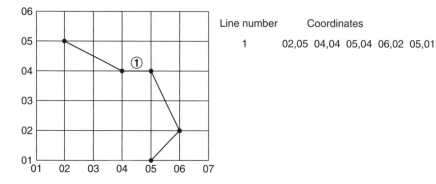

Line number	Coordinates
1	02,05 04,04 05,04 06,02 05,01

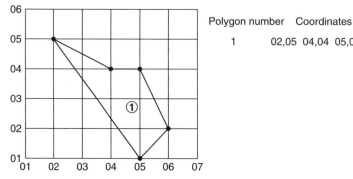

Polygon number	Coordinates
1	02,05 04,04 05,04 06,02 05,01 02,05

Figure 5.3   Points, lines and polygons

The points, lines or polygons are often referred to as 'spatial objects' or 'features' but, as suggested above, there is no other information about them and information about attributes is required. In a vector system, the attribute information is contained in a simple tabular database, as described in Box 5.1.

However, for the GIS to operate, the attribute database has to be linked to the spatial database, the link often being referred to as 'tagging'. The linkage of the two databases means that for any point, line or area, the appropriate attribute data can be accessed, often simply by pointing at a map created by the spatial data.

For example, an oil well might be represented on the spatial database as a point and geographically identified by a single coordinate pair. The attribute database would then include such information as the depth of the well, the chemical composition of the oil, the rate of extraction and any other relevant details.

Raster data models

Raster data models operate in an entirely different way. A 'raster' is an array of squares, known as pixels. Each pixel contains one item of attribute information and the arrangement of pixels in the raster identifies their locations, i.e. the raster itself is georeferenced. Pixels in the raster are identified by coordinate pairs but, unlike vector pairs, or OS maps, it is the pixels themselves that are identified, not the intersections of lines. Also, raster numbering starts at the top and works along the top line, then moves down to the second line. There are no separate spatial and attribute databases; both are contained in the same raster framework. A simplified example is shown in Figure 5.4. Since there is no separate attribute database, if different sets of attributes are required for the same area, a separate raster has to be created for each one. This may appear to be a drawback when compared to the attribute database in a vector model which can contain many different types of information but, in many ways, particularly for some kinds of statistical analysis, it is a distinct advantage.

The map in Figure 5.4 inevitably gives the impression that raster information is 'blocky', but rasters often consist of hundreds, perhaps thousands, of pixels. With huge numbers of pixels, each pixel is very small (at the scale at which they are usually viewed) and the impression of smooth curves is created (see Figure 5.7 on p. 90 for a demonstration of this).

Image data

A simple form of raster data is often referred to as an image. An image is digital data arranged in a raster but in which the individual pixels do not contain usable data values. Frequently used images include air photographs and paper maps which have been scanned in to a computer (which converts them to computer-readable digital form) and digital photographs. The pixels in these images usually represent just colours or shades of grey, rather than meaningful data which can be manipulated. In this respect, they are really only 'pictures' of the originals in digital form.

It is possible to use this non-numeric 'data' in GIS, often to good visual effect. For example, a scanned-in paper map has no numeric values, but it still makes an effective backdrop which can be overlain with other data created in different ways. Similarly, scanned air photographs or remotely sensed images are frequently used in the same way. Being in digital form means that images can be processed by the GIS in various ways. In particular, they can be resized or zoomed in and out. Most important, however, is that they can be georeferenced (i.e. georeferencing can be added to them). Once georeferenced, images can be used for measurement and further GIS information can be added to them in the

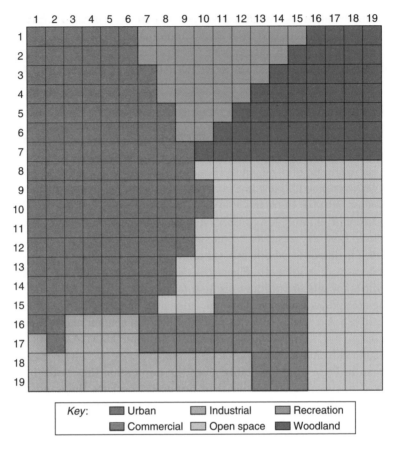

Figure 5.4   Raster diagram

form of vector layers. That is layers created in the vector format of points, lines and areas superimposed on the image. Layers are explained more fully later.

For these sorts of reasons, images are often used as backdrops for other kinds of information. For example, vector layers showing the alternative routes for proposed new roads could be displayed on top of a scanned image of a paper map or air photograph. Figure 5.11 (p. 97) shows an air photograph of the western edge of Heathrow airport and part of the M25, used as a base for information added by a GIS as part of the SAUL Project (pp. 96–98).

Layers

One major facility of vector GIS is that they can display the data, or, more usually, subsets of the data, in a variety of different ways, including various types of map. Each of these sets of data can be displayed separately and are often referred to as 'layers' of information about the same area. Any combination of layers can be superimposed on each other and output as a single layer. The output does not have to be visual but very frequently it is displayed as a map. Raster datasets also

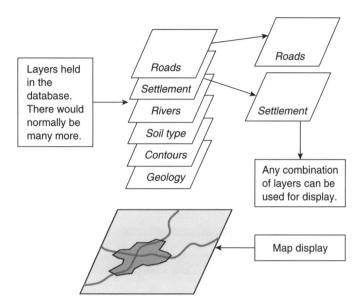

Figure 5.5   Layers diagram

contain layers but they cannot be displayed superimposed. However, the different layers can be used (rather than displayed) in combination for statistical analysis.

Layers can be turned on and off. The diagram in Figure 5.5 shows many layers held in the database but just two being extracted to create a map. The map in Figure 5.6, p. 88 has over 30 layers which can be switched on or off by clicking a box: the maps displayed demonstrate just three layers displayed in various combinations. Also, layers are not simply a device for viewing an area in a variety of ways. The use of layers enables GIS to perform some very sophisticated processing, some of the more important of which are described later, on pp. 89–90.

Data analysis

Another function (which usually comes before map display) is to analyse the data. This analysis is often of a mathematical nature and converts the raw data into usable information.

Numerical analysis

The data held in GIS is for the most part numeric data, and it is such data values which can be manipulated by the GIS. There is not space here to elaborate on all the numerical functions but three simple examples which illustrate these functions follow.

Measurement

The very fact that GIS data is georeferenced means that distance is easily measured and output in any measurement units, and either in tabular or map form.

88 UNDERSTANDING ENVIRONMENTAL ISSUES

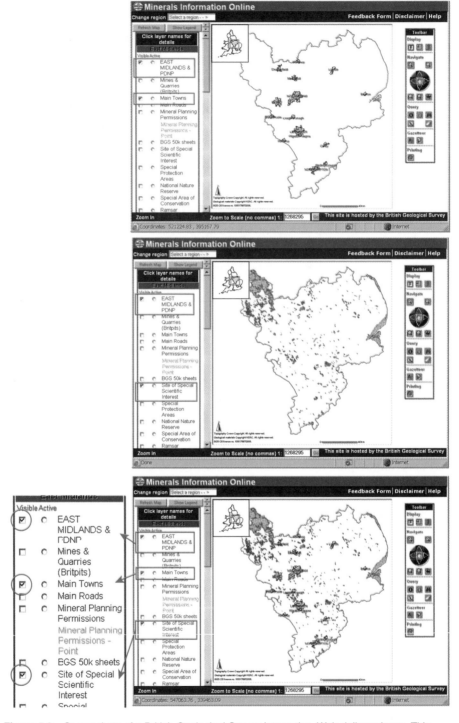

Figure 5.6 Screenshots of a British Geological Survey interactive, Web-delivered map. This map has over 30 layers which can be displayed by clicking the boxes to the left of the map.

Source: Crown/NERC

This might not seem too difficult – why not just take a ruler to a map? But since the geographical location of all the points in a vector model of an area, or all the pixels in a raster, are stored, GIS are able to calculate distances which are not straight (along roads or rivers, for example). Similarly, perimeters and areas can be calculated, as can 'buffers'. Buffers are zones within a given distance of a feature. For example, in planning for environmentally sensitive areas (ESAs), it may be helpful to prohibit any construction within a certain distance of, say, a lake, or river. Just imagine how difficult this would be to do on a paper map, measuring distances around every wriggle in the shape of the object.

Database query

Another form of analysis is database interrogation or query. Such queries involve identifying links between the attribute and the spatial databases.

There are two, so-called classic, GIS queries which are:

1 Query by location: what attributes can I find at this location?

In this case, a location is specified (often simply by pointing with a mouse) and the requested attributes are displayed, often as a map but equally often as a table of data. For example, pointing at a building on a large-scale map might produce extracts from a table showing the function of the building, its ground area and number of people employed.

2 Query by attribute: where can I find these attributes?

Such a query might request, 'Where can I find areas of clay soil?' and, usually, a map is displayed showing clay soils. This query may be executed simply by highlighting cells in a table.

Overlay

Overlay is a process used in many ways, but typically to derive a new area by overlaying the information from two existing ones, for example, to identify an area which is common to both. This process is not necessarily visual. In the case of raster data, two different data-sets are selected, and may be displayed, but separately (since you cannot 'see through' raster maps). Visual display is not, however, necessary.

Computations are then carried out, pixel by pixel, between the two data-sets to identify whether the values of the two pixels in the same location in each raster are the same or different. Clearly for this operation, it is a condition that the two data-sets are identical to each other in the number of rows and columns.

The pixels which are the same can then be displayed as a map, but if the objective is numerical (e.g. to discover the area in hectares), it may simply be output as a number. Figure 5.7 (p. 90) shows an example of overlay analysis.

Vector overlay is much more computationally complex. The identification of common areas is similar but the information about the new areas is different. Users need to identify which of the two is more useful for their intended purposes. For the technically competent, the two methods are described in some detail by Heywood et al. (2006) and Longley et al. (2005a).

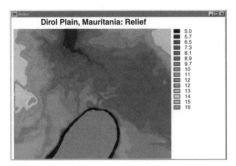

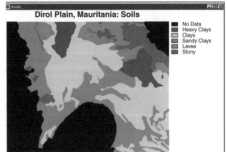

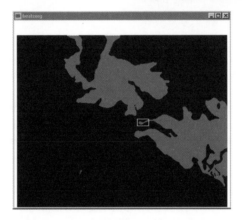

Figure 5.7 Raster overlay example. This example is taken from the Idrisi tutorial exercises. Idrisi is a raster-based GIS package (© Clarks Labs, Clark University)

The exercise is to improve the production of a crop called 'sorghum' by methods within the experience of the farmers themselves, rather than utilising expensive, environmentally damaging, externally imposed, technological solutions. The first step is to identify areas which are *both* floodable (the blue shades on the relief map) *and* which are clay soil (the yellow areas on the soils map). The overlay process, which involves the mathematical comparison of the data values of each individual pixel on the two maps, identifies the areas in blue on the third map.

As a by-product, these maps can demonstrate that lines and edges on raster maps appear as smooth curves because of the large number of pixels. A greatly enlarged section of the third map (in the yellow box) shows that the edges are actually the edges of pixels.

Source: Eastman, 1997

Spatial analysis

This is really an extended form of numerical analysis which attempts to derive models (usually with a mathematical basis), explaining processes in which geographical space is an important element. Although work in this area began much earlier, for example, with the well known models of Central Place Theory (Christaller, 1933) and the concentric growth of cities (Burgess, 1925), spatial analysis really took off in the 1950s and 1960s with the 'quantitative revolution' in geography. Christaller and Burgess were 'rediscovered' and many new spatial models were developed. However, the tools and models of spatial analysis appeared to be too simplistic to accommodate the variety and irrationality of human behaviour, and it was largely superseded by other geographical approaches.

More recently, the availability of huge amounts of digital data, ever increasing computational power and the spatial capabilities of GIS have given spatial analysis a new lease of life, even though many difficulties and widely varying viewpoints persist (Longley et al., 2005b). Some examples of spatial analytical techniques will be given later, in the section on decision making.

The third dimension

In addition to all that has been considered so far, GIS have powerful capabilities in manipulating and displaying elevation data. Note, again here, that it is *data* that is manipulated, not maps. Of course, maps and other forms of visual display are still very important but it is the elevation data which is used to create the displays.

At its simplest, a three-dimensional, i.e. perspective, view of an area can be produced, which, in itself, assists visual interpretation. One step further on is to 'drape' an image on to the three-dimensional view. The image might be an air photograph or a map. With this addition, the effectiveness of the view for interpretation is greatly increased. Effective use of a perspective view is shown in Figure 5.9 on p. 92.

Elevation data are also used to derive 'viewsheds' or to perform 'intervisibility analysis'. These show areas which are visible from a chosen viewpoint or identify areas from which something is visible. This exercise is easily done for a single line of sight using a contour map. A cross-sectional profile is drawn from the chosen point, using contour lines. Figure 5.8 illustrates the idea.

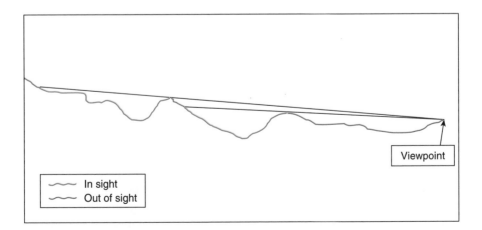

Viewpoint

In sight
Out of sight

Figure 5.8   Line of sight profile

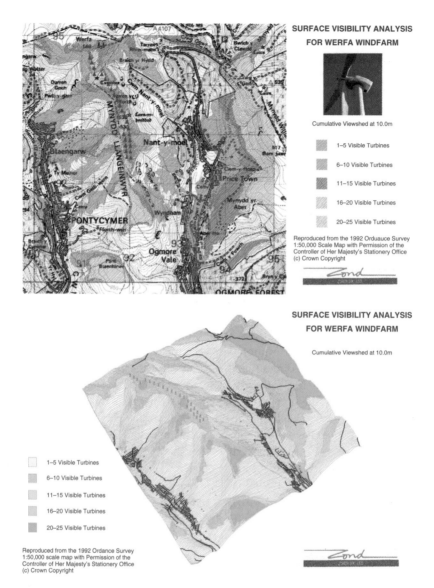

Figure 5.9 Surface visibility analysis of Werfa Windfarm. The map shows visible turbines as indicated by the key. The perspective view shows the same information in a way which is easier to assimilate visually

Source: Sparkes, A. and Kidner, D. (2006) *A GIS for the Environmental Impact of Wind Farms.* Reproduced from the 1992 Ordance Survey 1:50,000 scale map with permission of the Controller of Her Majesty's Stationery Office © Crown Copyright.

However, this is just a single line of sight. A viewshed calculates *all* areas visible from the chosen point, an operation which is impossible manually.

Figure 5.9 is an illustration of an intervisibility analysis to identify areas from which turbines in a windfarm are visible. As always, it is important to remember that viewsheds, slope angle and perspective views are all different outputs which are calculated from the same data.

In addition to the perspective view, the angle of slopes can be calculated from the data. Slope is a very useful criterion when considering sites for construction purposes.

At this point, it is appropriate to suggest how some of the capabilities of GIS, indicated above, might be used in a typical case (see Box 5.2).

Box 5.2   A hypothetical example: to determine suitable sites for a landfill waste disposal facility

The criteria are as follows:

1   Suitable surface geology – prevent leaching into groundwater. Query geology database (i.e. digital geology map) to extract suitable geology.
2   Further than one mile from existing settlement boundary. Query land-use data to identify settlement (extract settlement layer). Create one-mile buffer around each settlement periphery.
3   Overlay the results of 1 and 2 to identify areas common to both (i.e. suitable geology and outside buffers around each settlement).
4   Closer than a half mile from existing main road. Extract main roads layer. Create a half-mile buffer.
5   Overlay the outputs from 3 and 4 to identify areas common to both (i.e. the output from 3 which is also within the roads buffer).
6   Closer than a quarter mile from suitable surface drainage for treated effluent disposal. Extract surface drainage layer. Create a quarter-mile buffer.
7   Overlay the outputs from 5 and 6 to identify areas common to both.
8   Several potential sites might now be identified.
9   Create a viewshed from each potential site.
10 Overlay the outputs from 8 and 9. Eliminate sites in view of existing settlements.

Chapter 7, on waste, deals extensively with landfill and the problems of finding suitable sites.

GIS as a decision-making aid

The GIS processes described above indicate that GIS is particularly valuable in assisting decision makers. It is not, nor should it be, the only tool used. In any case, all decisions are made in the context of wider organisational, social, political and economic considerations, as is made clear throughout this book.

The hypothetical landfill example clearly sets out to be a decision-making process – that is its purpose. However, one of the problems with overlay techniques is that each layer is regarded as equally important in arriving at the decision and that is not always the case. It may be (and probably is) that visibility from a settlement is less important than suitable geology. A solution to this is multi-criteria evaluation (MCE) in raster-based GIS. This permits different criteria to be weighted according to their perceived importance by the decision makers. An extension of MCE is that different decision makers, who have differing opinions, are enabled to input their own weightings which the GIS is able to combine.

A very good interactive illustration of this capability is the House Hunting Game from the GeographyCal project. Most university networks should have GeographyCal available on them but, if not, there is a detailed description, with illustrations, in Heywood et al. (2006). Schuurman (2004: 107–10), also includes a case study of a Canadian landfill proposal to illustrate MCE, which makes a useful comparison with the hypothetical case above.

Another difficulty which affects decision making is the problem of definitions. It is self-evident that soil types or, say, wetlands do not have clear-cut boundaries – they change gradually from one type to another. Yet such boundaries appear on maps and in GIS data and decisions may be made on them. This problem is partly related to scale. The boundaries of settlements, for example, referred to above, may fairly be represented as lines at a fairly small scale, but at larger scales it is more difficult.

A further problem relates to differences of opinion between individuals or organisations in determining boundaries. How are the boundaries of, say, ESAs or SSSIs determined? The answer is almost certainly by economic and political as well as scientific argument. In effect, a GIS can help only up to a point in enlightening decision making.

The foregoing has constituted a brief résumé of some of the main capabilities of GIS, with some indications of applicability to environmental issues. It is now time to turn to some of the limitations of, and problems associated with, GIS which were mentioned briefly in the introduction. The main purpose of this section is to draw attention to the need for care in the use of GIS, or unquestioning acceptance of GIS output, neither of which is to deny the undoubted value and widespread use of GIS in relation to the environment.

Problems with GIS

During the 1990s, as GIS use was becoming widespread, criticisms began to emerge about the nature of GIS, the societal implications of its use and ethical problems related to these. The 'nature' of GIS refers to what they are, their modes of operating and, most important, the philosophical basis on which their validity for revealing 'truths' rest.

The main problem is that the use of GIS incorporates the method of enquiry used in the natural sciences, sometimes called 'the scientific method', though not everyone would agree that there is only one. Social scientists, which includes human geographers, have long been concerned that this approach is unable to cope with the complexity of human cultural, economic and social situations. In particular, spatial analysis (pp. 90–91) was subject to criticism before the advent of GIS, and some see GIS as providing a new legitimacy for an approach previously discredited among social scientists. The GIS community has not accepted these criticisms without response and the arguments have been extensively developed, notably by Pickles (1995), and reviewed more recently by Schuurman (2004) and Pickles (2005).

A related problem is the representation of the real world by models (mentioned earlier on p. 81). This problem was also referred to in the section on GIS in decision making, on pp. 93–94. In his excellent book, *How to Lie with Maps*, Mark Monmonier (1996: 74–81) includes a section on the preparation of an environmental assessment for a typical development project application in the USA. In it, he shows how there are many ways in which, through maps (which are, of course, models of the real world), the assessment can be made to reflect the views of one or other of the parties involved in the application. These include such things as scale (small but potentially important details may not appear in small-scale maps), the definition of boundaries (of, for example, floodplains), the use of inappropriate input data, and several others, including the use of GIS. Of course, these manipulations are not actually 'lies'. The possibility of presenting information in favour of a particular point of view rests on the problematic nature of representation. It is not possible to display reality in its entirety and the way in which maps (and GIS) represent reality is always subjective, despite the appearance of objectivity.

These problems extend into the realm of ethics. Throughout the developed world, there are organisations which could, in some way, be seen as invading privacy or undertaking some form of surveillance, or acting on behalf of people, all without their knowledge. The ubiquitous supermarket loyalty card is one good example, which provides the supermarket with information, not just about spending habits but about demographic and spatial information also. Pickles (1995) and Curry (1995), elaborate on these issues and refer to a great many examples.

There is also the issue of transparency and accountability. Even if organisations are not seeking to invade privacy or to exercise some kind of control, and even if they are seen to be beneficial (such, perhaps, as environmental agencies), they may operate in a way which excludes those for whose benefit they operate. They may think of themselves as the experts, using technology which the public cannot understand and of being able to produce the right answer because the technology is so sophisticated. But, of course, the technology can be fallible (the GIGO principle, p. 80) and the data-selection and data-processing methodology is inevitably subjective, as we have seen. This is not to suggest that technology cannot help, but that the operators need to be aware of the limitations and need to make their beneficiaries aware of these considerations.

This section may be perceived as negative but it is not. It is very important that the limitations of technology and the societal implications of its use are understood, but this is not to imply that GIS and related technologies are invalid. Indeed, this chapter would not have been written if this were the case. Although the issues have not disappeared, there is less polarisation now than previously between social scientists and the GIS communities, and greater fruitful dialogue between them (Pickles, 2005; Schuurman, 2004).

Public participatory GIS

Some of the foregoing criticisms suggest that GIS was associated with techno-cratic elitism and a paternalistic view of the functioning of public bodies. Partly in response to this view, and partly in relation to the growing acceptance of public participation in decision making, there emerged a movement to demystify GIS and to employ its capabilities in the empowerment of ordinary people. This became known as 'public participatory GIS' (PPGIS), or one of several similar names. The key to PPGIS is the participatory element rather than the GIS, which is secondary to it (McCall, 2004). PPGIS is now widely used in natural resource planning and land-use management (Craig et al., 2002), and is eminently suitable to be employed in environmental issues. It is important to see PPGIS as a method to complement existing methods of public participation, and it usually takes place within a larger project with other activities taking place around it. A case study of such a project follows.

Case study: Groundwork Thames Valley (GTV) Colne Valley Project in association with Sustainable Accessible Urban Landscapes (SAUL)

Groundwork Thames Valley is an environmental charity using PPGIS techniques to undertake community consultation and planning. SAUL is a transnational part-nership project part-funded by the European Union.

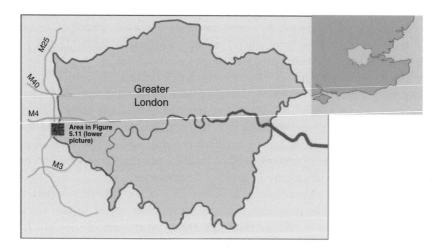

Figure 5.10 The location of the study area in SE England

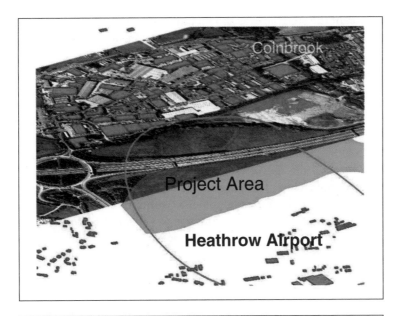

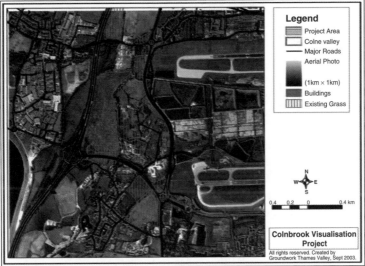

Figure 5.11 GTV SAUL project area
Source: Groundwork Thames Valley

The Colne Valley Regional Park was established to capitalise on the potential recreational value of a diverse landscape of water and woodland. Recently though, the strategy has been dominated at its southern end by urban encroachment from controversial development schemes, such as Heathrow Terminal 5. The absence of positive practices has resulted in illegal and temporary land uses that create an impression of dereliction and neglect. Upon completion of Terminal 5, the adjacent area will become available for regeneration. The project focuses on this area and the local community of Colnbrook.

The location of the study area is shown in Figure 5.10 and the details are shown in Figure 5.11, which superimposes vector layers on image backgrounds (aerial photographs), as described on pp. 85–86.

The role of Groundwork Thames Valley is to ensure that the community of Colnbrook is involved in the planning processes for this area by community participation workshops and by the development of a virtual reality landscape project through which local people can get involved, particularly those often marginalized or under-represented (ethnic minorities, disabled, young people and children). Creating visualisations (shown in Figures 5.1 and 5.2) and depicting information through the use of GIS is a way of engaging with local people and explaining sometimes complex ideas in a simple form, and hence empowering them to take responsibility for developments within their neighbourhoods.

As this chapter began, with an introduction to visualisations created by various technologies, so this part of it ends with the description of the uses of those same visualisations in PPGIS. The next section of the chapter moves on to consider another key technology: remote sensing.

Introduction to Remote Sensing

Environmental scientists are concerned with observing natural phenomena, making careful observations and measurements of these phenomena, and then using these data to test hypotheses, with the goal of understanding the nature of the object or process under investigation. Usually data are acquired *in situ*, or via laboratory analysis of samples taken from the field. In either case, the measurements and observations are made by, in one way or another, directly touching the object of interest. For example, the dissolved phosphate content of a soil can be determined by sampling the soil, returning the samples to a laboratory, extracting the phosphate from the soil, and determining the concentration of phosphate in the extractant using a spectrophotometer. The acidity of the water in a lake can be determined by suspending a pH electrode from a boat; the concentration of nitrogen dioxide in the atmosphere, an important traffic-related pollutant, can be measured by placing diffusion tubes close to roads.

Much of the data used by environmental scientists is collected in this way. However, it is often possible, occasionally desirable, and sometimes necessary, to perform such measurements from a distance. Very generally, remote sensing, sometimes called 'Earth Observation Science' can be defined as the science of measuring the properties of an object without touching it. But what does that mean, how do we do it, and of what relevance is it to environmental scientists?

What is remote sensing?

Environmental remote sensing is concerned with making observations and measurements of the physical, chemical and biological properties of Earth's surface and atmosphere without direct physical contact. This information is obtained by measuring the amount of electromagnetic radiation (EMR) (see Box 5.3) reflected or emitted from the materials and objects that cover that surface.

Box 5.3   Electromagnetic radiation

Electromagnetic energy (EMR; often referred to as light, radiation, or simply energy) is generated whenever charged particles are accelerated. It is produced by any object with a temperature greater than absolute zero (0 Kelvin, or –273.15 °C).

EMR is complicated, but at its simplest can be considered as a stream of photons travelling in a wave-like pattern, where each photon carries a certain amount of energy. In a vacuum these bundles of energy (also called quanta) move at the speed of light. EMR is actually made up of two force-fields, moving at right angles to each other, one electric, the other magnetic, below (a).

Bodies such as the Sun and Earth emit a continuous spectrum of energy. This means that they generate EMR over a contiguous range of wavelengths. The curves below (b) show how the spectral composition of the energy emitted by a body (i.e. how much it emits at a given wavelength) changes as a function of temperature. The uppermost

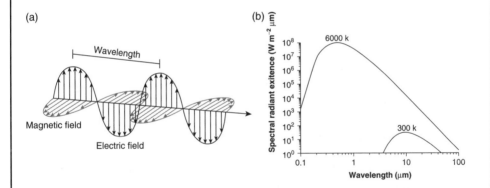

curve shows the spectrum of energy emitted by a blackbody at 6000 K, approximately equal to the surface temperature of the Sun (a blackbody is something that emits energy at the maximum rate possible, given its internal temperature). The lower curve shows the same quantity, calculated for a 300 K blackbody (about equal to the average surface temperature of Earth). Clearly, the Sun emits much more energy at all wavelengths than the Earth, the total energy being given by the area under their respective curves. You will also notice that the wavelength at which the Sun and Earth emit most energy differs. The Sun emits most EMR at a wavelength of about 0.5 μm. The Earth is not hot enough to emit energy at these short wavelengths; it emits most of its energy at about 9 μm.

Another way to say that emittance from the Sun reaches a peak between 0.45 and 0.70 μm is to say that the Sun emits most of its energy in the visible region of the electromagnetic spectrum (below). The EM spectrum is an attempt to classify, or group, EMR according to

(Continued)

its wavelength. The visible part of the EM spectrum is so called because it corresponds to the wavelengths of light to which the human eye is sensitive. EMR with wavelengths of between 0.70 and 1.10 μm is often referred to as infrared radiation, whereas EMR with wavelengths between about 0.01 and 0.38 μm is referred to as ultraviolet light.

Although we talk about different "types" of EMR, they are in fact all the same thing: energy. The only difference between microwaves and gamma rays is the amount of energy carried by the photons, which determines the wavelength (the physical distance between the peaks of the waveform) and frequency (the number of waveforms passing a fixed point in space during a given period of time, usually one second) of the waves themselves. In environmental remote sensing it is most common, however, to distinguish EMR on the basis of its wavelength.

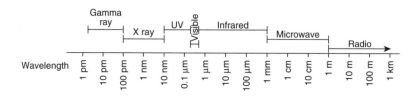

The human eye and brain combine to form a very effective organic remote sensing system. Parts of this paper have been treated with a pigment that absorbs light. As a result, light is reflected in a particular pattern when the page is illuminated; areas occupied by the letters and numbers absorb all the incident light, and appear black, while adjacent areas of the page reflect all of the incident light, and appear white. The light that is reflected enters your eye via the pupil and is focused onto the retina to form an image of the page. This image, which is simply a map of where light is being reflected from the paper (or, conversely, where light is being absorbed), is recorded by millions of photo-receptors distributed across the retina. These receptors contain pigments which, when illuminated, absorb light and convert it via a chemical reaction into a series of nerve impulses. As the light reflected from different parts of the page varies according to the particular pattern in which it has been dyed, so the image of the page on the retina causes these receptors to be turned on in a particular pattern, resulting in the transmission of a set of electrical impulses to the brain, via the optic nerve. In this way, your eyes have acquired, remotely, a map of how much EMR this piece of paper reflects. Your brain recognises that the variations in reflectivity correspond to a collection of letters, and has interpreted the permutations of letters as words, and the permutations of words as sentences. Hopefully, they make sense. Although environmental remote sensing relies on the use of artificial eyes often carried on satellites orbiting hundreds of kilometres above the Earth's surface, the principles by which they acquire information regarding our environment are largely the same (see Box 5.4).

Box 5.4   How does remote sensing work?

Most remote sensing uses energy provided by the Sun (1) in the diagram below. This energy travels through space unhindered. Upon entering Earth's atmosphere, some of it is scattered and absorbed by the gases and particles within the atmosphere (2), failing to make it to the surface. However, there are certain wavelengths of EMR to which the atmosphere is relatively transparent and this EMR reaches the surface (these wavelengths, where the transmissivity of the atmosphere approaches 100%, are called atmospheric windows, and are shown below).

The energy that makes it through the atmosphere is then free to interact with the materials (i.e. vegetation, rocks, soils, etc.) that comprise the Earth's surface (3). The energy that strikes these materials is said to be partitioned, and can be either a) reflected back away from the surface, b) absorbed by the surface, later to be emitted as thermal energy, or c) transmitted through the surface (although as most natural materials are opaque, transmission is often ignored). If a surface absorbs a lot of EMR at a particular wavelength, it must reflect relatively little at that wavelength; if it absorbs only a little, it must reflect a lot.

Because different Earth surface materials are made of different chemical elements, combined in different ways, with different physical structures, they absorb and reflect different amounts of energy at different wavelengths. As a result, the spectral composition of the EMR that is reflected is changed during the aforementioned interaction (see Box 5.5). By using instruments that are designed to measure the spectral composition of the reflected energy (i.e. how much is reflected back at each wavelength), we can distinguish different materials on the Earth's surface.

The instruments used are referred to as sensors, which are mounted on platforms. The platform may be an artificial satellite orbiting 700 km above the Earth's surface, but could equally be an aeroplane. 'How remote is remote?' is a bit like asking, 'How long is a piece of string?', although for reasons discussed in the text, most environmental remote sensing takes place from Earth's orbit.

The sensor measures the amount of EMR reflected (4) and converts this into a digital format (see Box 5.6). This information is then transmitted back to Earth (a process called telemetry (5) where it is archived, either on CDs, DVDs or, increasingly, on computer hard disks.

The final step in the chain (6) comes when the remote sensing data arrive on the desk of the analyst. At this point, the satellite images must be processed (see Box 5.6) and interpreted, preferably by a scientist with knowledge of remote sensing, image processing, and also the area of the Earth's surface to which the image pertains.

Although most remote sensing uses the Sun as the source of EMR, we can also study variations in the amount of energy that the Earth emits. Both of these forms of remote sensing are termed 'passive' because the system relies on EMR generated naturally. 'Active' remote sensing involves using instruments that generate their own EMR, which is 'bounced' off the Earth's surface. By measuring the strength and properties of the energy that is backscattered, we can study how the Earth's surface materials interact with, for example, microwaves.

(Continued)

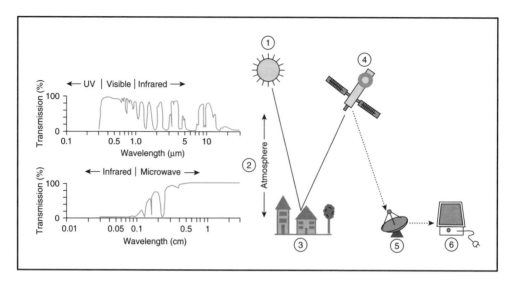

The human eye is sensitive only to EMR in the visible part of the EM spectrum. However, the strength of remote sensing lies in the fact that we can design and build artificial sensors that can 'see' in parts of the electromagnetic spectrum that are invisible to us. This is important because the materials that constitute our environment (e.g. different types of vegetation; different rock types) reflect and emit different amounts of EMR at different wavelengths (see Box 5.5). By studying the way in which radiation interacts with materials at other wavelengths, we can use remote sensing not only to tell different materials apart, but also to study their physical and chemical properties.

Box 5.5   Spectral signatures

The amount of energy reflected by a surface, and how this varies with wavelength, is referred to as its 'spectral response function', or more commonly its 'spectral signature'. The chart below shows the spectral signatures of some common Earth surface materials. The amount of EMR reflected is shown on the y-axis; wavelength is shown on the x-axis. At points where the spectral signatures of two materials cross over (for example, the spectral response curves of vegetation and soil intersect at 1.40 µm) the amount of EMR reflected from those materials is the same at that wavelength, and the materials are spectrally indistinguishable. Notice, however, that even in this narrow region of the electromagnetic spectrum there are many more wavelengths at which vegetation and soil reflect very different amounts of radiation and are, therefore, spectrally distinct.

The shape of the spectral response curves is determined by the physical and chemical 'make up' of the material. For example, the chlorophyll present in green leaves absorbs visible blue and red light, but not green. This causes a reflectance peak in the spectral signature of healthy vegetation at about 0.53 µm, the wavelength of visible green light, explaining why healthy plants appear green to our eyes. These, combined with absorption troughs and reflectance peaks related to leaf cell structure and moisture

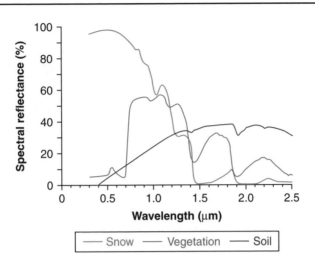

content at other wavelengths, combine to give green vegetation its own unique spectral 'fingerprint'.

By using remote sensing devices sensitive to EMR in the thermal infrared region of the electromagnetic spectrum (about 8 to 14 μm) we can also distinguish different materials on the basis of how much energy they emit at different wavelengths, using the same general principles.

The details of how much EMR is reflected or emitted from the Earth's surface, and how this varies as a function of wavelength, are stored digitally as a string of binary numbers. This means that not only can computers be used to display the information that the remote sensing data contain (see Box 5.6), but we can also perform quite complex mathematical transformations on the data to increase the amount of information we can extract. The relative ease with which remote sensing data can now be obtained via the Internet, and the proliferation of powerful desktop computers, has made remote sensing an increasingly accessible technology for environmental scientists.

BOX 5.6   Images, not photographs

Satellite images are often erroneously referred to as satellite photographs. Although photography is a kind of remote sensing, the images you see on the daily news weather bulletin are acquired in a very different way.

As the orbiting satellite sensor measures and records the amount of EMR reflected or emitted from the Earth's surface (see Box 5.4), it does so in a systematic way. Below to the right is part of a visible wavelength image of a scoria cone on Mount Etna volcano, in Sicily, acquired by the widely used Landsat Thematic Mapper (TM). Each square area you can see is

(Continued)

referred to as a pixel (a contraction of "picture element"). These pixels are equivalent to an area on the
ground of about 30 m × 30 m. Although the details of
the image acquisition process are beyond the scope of
this chapter, you can imagine that the Landsat TM sensor divides the Earth's surface into a series of roughly
square parcels (rather like a chess board) and then
makes a measurement of the EMR emanating from
each one. By doing this in a regular manner, a two-
dimensional image of the surface is produced.

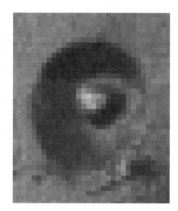

The amount of EMR received from each parcel (or
'ground resolution element'; GRE) is recorded by the
sensor as a 'digital number' or DN. As the DN is just a
number, satellite data lend themselves to display and analysis in a computer environment. The image to the right above was produced by telling the computer to colour pixels with a low DN black (i.e. to signify that little or no reflectance was detected from the
surface) and to colour pixels with a high DN white (i.e. to signify that a lot of reflectance
was detected from the surface), with pixels with intermediate DNs given an appropriate
shade of grey. In this way the computer can convert a series of DN (remotely sensed
data) into a picture (a remotely sensed image).

The real power of remote sensing as a geographical analysis technique comes
from the fact that satellite sensors can be designed to make measurements of Earth
surface reflectance in many different regions of the EM spectrum. Indeed, making
simultaneous measurements of the amount of EMR reflected (and/or emitted) from
each pixel is the only way that we can hope to resolve differences in the spectral signatures of the materials that these pixels contain. A Landsat TM image in fact contains
seven 'bands' of data, where each band contains reflectance/emittance information for
a different wavelength interval. For example, a TM band 1 image shows the portion
of the total energy reflected by the surface that has wavelengths of between 0.45 and
0.52 μm. TM provides not one but seven images of Earth's surface; three in the visible part of the EM spectrum, one in the near-infrared, two in the short-wave infrared,
and one in the long-wave infrared (below). Such an image is called a 'multi-spectral
image'.

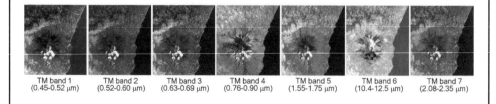

| TM band 1 | TM band 2 | TM band 3 | TM band 4 | TM band 5 | TM band 6 | TM band 7 |
| (0.45-0.52 μm) | (0.52-0.60 μm) | (0.63-0.69 μm) | (0.76-0.90 μm) | (1.55-1.75 μm) | (10.4-12.5 μm) | (2.08-2.35 μm) |

There are, and have been, many different Earth observation satellites in orbit,
carrying a wide range of sensors. Some, such as the Landsat Multi-spectral

Scanner (MSS), Thematic Mapper (TM), and Enhanced Thematic Mapper Plus (ETM+), have been dedicated to Earth resource applications, including agriculture, vegetation science, geology, and water quality. Other sensors, such as the Advanced Very High Resolution Radiometer (AVHRR), are designed to study Earth's atmosphere, oceans, and weather. As satellite sensors are designed with specific applications in mind, the nature of data provided by one is likely to be significantly different from that provided by another. These differences are usually quantified in terms of four 'resolutions': spatial, temporal, spectral, and radiometric (see Box 5.7).

Box 5.7   Properties of remotely sensed data

The properties of remote sensing data acquired by different imaging sensors are often compared in terms of four 'resolutions':

Spatial resolution: although spatial resolution is (potentially) a complex concept, at its simplest it can be considered to be the amount of spatial detail that is preserved in a remotely sensed image of an area on the ground. More formally, it is the minimum distance between two objects that a sensor can record distinctly. Spatial resolution is generally equivalent to (and is often stated as) the pixel size of the image. Below are two Landsat ETM+ images of London Heathrow airport. The image to the left has a spatial resolution of 15 m; the one to the right has a spatial resolution of 30 m. Notice how the 15 m data allow the runways to be resolved, in addition to roads in the residential areas surrounding the airport. This detail is not recorded in the 30 m data, which is said to have 'lower' or 'coarser' spatial resolution.

Temporal resolution: this is how often a sensor acquires an image for a particular point on the Earth's surface. Usually this is specified as the time that elapses between the

(Continued)

sensor acquiring successive images of a particular ground target from the same position in the sky. The Landsat 7 ETM+ sensor acquires an image of Heathrow airport once every 16 days. In contrast, the temporal resolution of the AVHRR sensor is 12 hours and the data it provides are, therefore, said to have 'higher' temporal resolution.

Radiometric resolution: this is the ability of the sensor to discriminate subtle differences in the EMR reflected or emitted by the surface. The Landsat ETM+ data shown above are 8-bit data. This means that the sensor can record one of 256 different DN for each pixel. The MODIS sensor acquires 12-bit data, allowing each pixel to have one of 4096 different DN. A sensor which uses more 'bits' (or DN) to record image brightness is said to have 'higher' radiometric resolution. Radiometric resolution is related to the precision of the measurement (but not its accuracy).

Spectral resolution: this is defined as the number and width of the spectral bands of the EM spectrum to which a remote sensing device is sensitive. In Box 5.4, you learnt that Landsat TM makes measurements of the amount of reflected and emitted EMR in seven different wavebands for each 30 × 30 m area on the Earth's surface. A sensor which records more wavebands of data would be described as having 'higher' spectral resolution, particularly if those wavebands were narrower. For example, the Earth Observing-1 Hyperion sensor has the same spatial resolution as Landsat TM (30 m) but acquires 242 wavebands of data. Such data are termed 'hyperspectral'. Clearly, higher spectral resolution provides more measurements of how the amount of EMR reflected (or emitted) from a surface varies as a function of wavelength. This, in turn, increases the likelihood that the data will be able to resolve differences in spectral signatures, thus increasing our ability to identify and discriminate between different Earth surface materials.

The question of which type of data is 'better' is difficult to answer, as this depends largely on the intended environmental application. For example, a scientist studying the environmental impact of invasive plant species will probably require high spatial resolution data (i.e. 30 m or better), because invasive and indigenous plants often intermingle over relatively short spatial scales. However, as the phenomenon does not progress at a rapid rate, it is probably sufficient to have access to one or two images per year to quantify the rate at which the process occurs (i.e. low temporal resolution). Meteorologists and climatologists have a different set of priorities. Variables such as sea surface and cloud-top temperatures do not vary significantly over such small spatial scales, and low spatial resolution data are perfectly adequate. Low temporal resolution data, on the other hand, may be useless (can you imagine receiving a weather forecast once every 16 days?) Environmental scientists studying dynamic process, such as wildfires and atmospheric circulation, require high temporal resolution data, often using images acquired once every 12 hours or even once every 15 minutes.

Why use remote sensing to study the environment?

There are a variety of reasons why remote sensing presents an attractive means to collect data pertaining to the world around us. Some environmental variables, for example, the temperature of the Earth's oceans, simply vary over too great an area

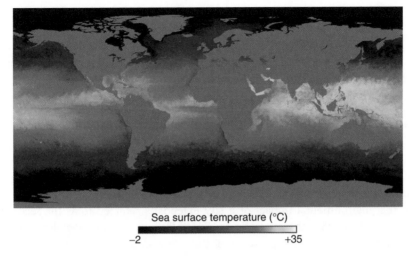

Sea surface temperature (°C)

−2   +35

Figure 5.12 Map of sea surface temperatures (SSTs) compiled by NASA's EOS Terra MODIS sensor. MODIS, which stands for Moderate Resolution Imaging Spectroradiometer, images the entire Earth four times per day using two near-identical sensors flown on two different satellites, allowing global maps of SST to be updated frequently. Image courtesy of NASA (http://visibleearth.nasa.gov)

to be adequately measured on the ground. A single satellite image, collected in a matter of seconds, can provide measurements of sea-surface temperature over hundreds to thousands of square kilometres, almost at an instant (Figure 5.12).

Furthermore, these data are collected frequently. Satellites are placed in orbits that allow them to acquire data for the same points on the Earth's surface at regular and repeated intervals. Although individual satellite sensors have limited lifetimes, government agencies responsible for Earth observation, such as the National Aeronautics and Space Administration (NASA) and the European Space Agency (ESA), have a strong commitment to data continuity, constantly launching new satellite sensors to replace old ones. In this way, decadal time-series of environmental data have been compiled. This is important when the environmental processes under investigation, for example tropical deforestation, occur at a relatively slow rate. A sensor mounted on a satellite not only makes repeated measurements, but the fact that the same sensor is used on each occasion results in a data-set of uniform quality (although a great deal of effort is diverted to calibrating and characterising the performance of the sensors to achieve this, to ensure that the data are 'accurate' as well as 'precise'). *In situ* data, possibly collected by different investigators using differing equipment and methodologies, may lack similar consistency, making long-term comparisons difficult.

A significant factor in the use of remote sensing in many environmental applications is cost. Although satellites themselves are quite expensive to design, build, launch and operate (the most recent of NASA's Landsat satellite series, Landsat 7, cost around US $700 million), the cost of the data to the end-user is low. A Landsat 7 Enhanced Thematic Mapper Plus (ETM+) image currently costs about $200, and covers an area of 35,000 square kilometres. In fact, remote sensing data-sets are increasingly becoming freely available via the Internet. When faced with the task of, for example, compiling a natural resource inventory at the continental scale, the ability of remote sensing to provide data of a uniform quality, covering large areas, at low overall cost, makes it an attractive option.

Hopefully, the preceding discussion has given you a basic idea of what remote sensing is. The final part of the chapter provides an introduction to how remote sensing can be used to provide information relating to our environment.

Remote sensing the environmental impact of urbanisation

Cities cover less than 2% of the Earth's surface, yet contain over half of the world's population, placing a demand for energy and resources on the area immediately sur-rounding the city that has far-reaching effects on the environment. The area over which the environmental impact of a city extends is referred to as the 'ecological footprint' and includes the area affected by pollution, resource development, and transportation caused by the presence of the city (Goudie, 2001 and see Box 1.1 in this book's Introduction).

Urbanisation modifies land cover via the replacement of soil and vegetation, with a combination of, among other things, concrete, asphalt, metal, and glass. This tran-sition from pervious to impervious cover type is a widely accepted indicator of urban-isation, and has profound effects on the pre-existing hydrology and microclimate. By concentrating resources and productivity, cities are also major sources of pollution, including atmospheric pollution from car exhausts, water pollution caused by urban run-off, and excesses of heat, light, and noise. Of all the impacts that urbanisation can have on the environment, two in particular have been extensively studied using remote sensing techniques: urban sprawl and the urban heat island effect.

At its most general, urban sprawl can be thought of as the pattern of land consumption that results from the urbanisation and suburbanisation of the land-scape (Wilson et al., 2003), and is considered to be a problem by environmental man-agers for a variety of reasons. A positive feedback exists between urban sprawl, the consumption of fossil fuels and emissions of greenhouse gases because, as the city becomes larger, cars become more 'essential'. Sprawl has also been credited with a range of negative socio-economic impacts including increased commute times (which reduce the time people spend both at work and at home), reduced social interaction at the community level, and the proliferation of out-of-town shopping malls at the expense of local centres of commerce (Ewing, 1997; Pedersen, et al., 1999; Wilson et al., 2003).

Remote sensing has been used to monitor and quantify urban sprawl using a vari-ety of methods. The replacement of soil and vegetation by asphalt and concrete is easily resolved as a change in the spectral signature of the land surface. Figure 5.13 depicts a pair of Landsat images that show the city of Las Vegas, in the Nevada desert, acquired in 1972 and 2001, in which the expansion of the city is easy to identify.

During the night, the Earth is much too cool to emit significant radiation at visible wavelengths. However, artificial sources of light, such as street lights, can be seen in night-time, visible-wavelength, satellite images (Figure 5.14). Satellite-mapped variations in both the areal extent and intensity of these lights have been used as a proxy for delineating the extent of urban areas and, indirectly, estimat-ing population densities (e.g. Imhoff, et al., 1997; Sutton, 2003). In this image,

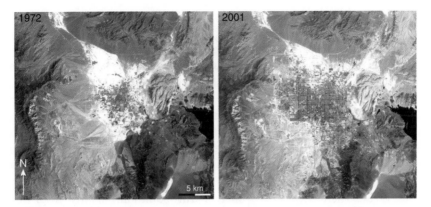

Figure 5.13 Two satellite images of Las Vegas, Nevada. The image on the left was acquired by the Landsat Multi-Spectral Scanner in 1972; the image on the right was acquired by the Landsat Enhanced Thematic Mapper Plus in 2001. The images have been processed so that healthy vegetation appears red

Figure 5.14 Visible-wavelength image showing the distribution of stable lights in western Europe during the period 1994 to 1995. This image was compiled from several images acquired by the US Department of Defense, Defense Meteorological Satellite Program. Image and data processing by NOAA's National Geophysical Data Center. DMSP data collected by the US Air Force Weather Agency

population centres are conspicuous either by their presence (e.g. Madrid, Paris and London) or absence (e.g. North Africa). Lights in the vicinity of the North Sea correspond to the flaring of natural gas at oil and gas platforms.

Not all urbanisation constitutes urban sprawl. Indeed, some kinds of urban growth, such as the infilling of derelict land in existing urban areas, act in direct opposition to sprawl per se. Similarly cities do not grow simply by expanding their outer perimeters, with growth often occurring in a linear pattern (i.e. along major transport routes) or like the branches of a tree. The repeat observation capability of remote sensing satellites provides a means not only to detect the change from non-urban (or undeveloped) to urban (or developed) land, but also to identify patterns in this change that can help urban planners understand *how* cities grow (e.g. Wilson et al., 2003).

Cities create their own microclimates. Air pollution, mainly from vehicle exhausts, modifies the composition of the urban atmosphere. The replacement of natural surface cover types with artificial materials alters the energy budget of the surface, by drastically changing its albedo and the amount of surface moisture available for evapo-transpiration, while construction of buildings and other structures modifies near-surface air–flow (i.e. winds; Dousset and Gourmelon, 2003). One of the most obvious results of this climate modification is the propensity of cities to have higher air and surface temperatures than the rural areas surrounding them, a phenomenon widely known as the 'urban heat island' effect (UHI; Oke, 1973). Increased temperatures in urban areas lead to an increased propensity for the development of smog and ground-level ozone, with obvious implications for human health. Higher temperatures also increase the rate at which cities consume energy (and therefore fossil fuels) via the increased use of air conditioning in commercial and residential buildings (Rosenfeld, et al., 1995).

Of course, the UHI phenomenon was recognised long before the advent of satellite remote sensing, and extensive records of urban surface temperature measurements, compiled from weather station measurements and observations made from car transects, exist. However, the urban heat island effect has been increasingly studied using satellite remote sensing (see Voogt and Oke, 2003 for a review), which provides improved spatial and temporal resolution, albeit with a shorter observation record. Measuring the UHI involves using thermal infrared images to compare the average surface temperature of a city with its surroundings (e.g. Streutker, 2003). Figure 5.15 shows a night-time AVHRR image of southern England in which the urban heat island created by several cities is apparent.

Remote sensing global wildfires

'Wildfire' is the term given to the uncontrolled burning of forests, grasslands, and brushlands. Wildfires can occur anywhere there is a source of fuel, a source of ignition, and weather conditions appropriate for the fuel to burn (Smith, 2001). They occur most frequently in the Mediterranean and continental climatic zones, where periods of vegetation growth are followed by extended periods of drought, high temperatures, and sustained dry winds, which serve to expel moisture from the vegetation, a precondition for pyrolysis and combustion. Wildfires can start naturally as a result of lightning strikes but are increasingly caused by humans,

Figure 5.15 Night-time thermal image of part of the British Isles. The image has been processed so that cold surfaces appear dark and warmer surfaces appear bright. As this is a night-time image, the ocean is warmer (and therefore brighter) than the adjacent land. Urban heat islands generated by several major UK cities are indicated

either accidentally (e.g. mismanaged agricultural burns, unruly campfires), or deliberately (i.e. arson).

Wildfires facilitate many essential ecosystem functions, including seed release, germination, and mineral recycling. However, the expansion of the world's cities means that an increasing number of people live at the urban–wildland interface, making wildfires an increasingly significant natural hazard. Although loss of life is relatively rare, destruction of property and infrastructure is not. Large wildfires, such as those in Indonesia in 1997–8 also affect human health by loading the atmosphere with gaseous and particulate aerosols (e.g. Sutherland et al., 2005). Understanding the geography of wildfires and their effects is of importance to a wide range of environmental researchers.

Wildfires are clearly dangerous to approach. They also occur over large, and sometimes remote, geographical areas, making remote sensing an essential tool in their analysis. It has been demonstrated that remote sensing can contribute to the understanding of wildfires and their effects by a) detecting the presence of wildfires and monitoring their distribution in space and time (e.g. Kaufman et al., 1998); b) predicting those areas which are *likely* to be at risk of wildfires *before* they occur (e.g. Roberts et al., 2003); and c) quantifying the gaseous and particulate atmospheric pollution that results from wildfire events (e.g. Roberts et al., 2005).

Wildfires are, obviously, hot, and temperatures as high as 1800 K have been observed for the most intense crown fires. As the temperature of a body increases,

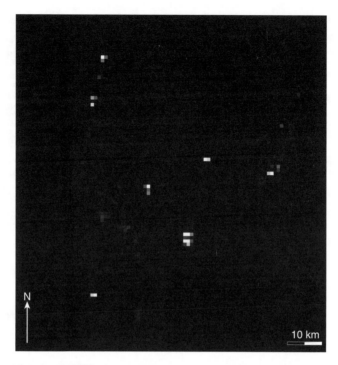

Figure 5.16 Night-time MODIS image of vegetation fires in central Africa. The spatial resolution of this MODIS image, which shows how much EMR the Earth was emitting at 3.959 μm, is 1 km. The white pixels contain fires, which although much smaller than the image pixels emit sufficient amounts of energy in this region of the EM spectrum to stand out

not only does it radiate increasing amounts of energy at all wavelengths, but the wavelength at which it emits most of its energy shifts to increasingly shorter wavelengths (this is what causes molten metal to 'glow' red; if heated further, the metal can become so hot that sufficient light is emitted with green and blue wavelengths for the metal to appear 'white-hot'). The same principles are true for wildfires, making them relatively easy to detect in satellite images, provided the satellite 'looks' at the appropriate wavelengths. Figure 5.16 shows part of an image of central Africa acquired by a satellite sensor called MODIS, which contains a group of wildfires. The spatial resolution of the image is 1 km, much larger than the size of the actual fires themselves. However, because the fires emit so much energy at this middle-infrared wavelength they cause the pixel in which they are contained to appear much brighter than the surrounding pixels that do not contain fires. By examining a succession of images for such differences in emitted energy, wildfires can be detected and monitored.

However, prevention is better than cure, and remote sensing can also be used to help predict where wildfires are most likely to occur, thus allowing fire mitigation resources to be targeted more economically. As previously mentioned, the nature and state of the fuel (i.e. vegetation) is of great importance in determining the level of wildfire risk within a particular area. Remote sensing can be used to distinguish the different types of fuel (e.g. some species of vegetation, such as eucalyptus, are more combustible than others because they contain flammable oils), and also to determine its relative moisture content. Satellite data can also provide estimates of the amount of biomass available for combustion and show whether the vegetation is live or dead. By mapping these types of variables, remote sensing can be used to provide data for input into GIS systems, allowing areas at high risk of wildfires to be identified.

Cumulatively, wildfires are significant sources of atmospheric pollution, via the emission of trace gases and aerosols (small particles) during combustion. Many of these gases are capable of absorbing infrared energy (Andreae and Merlet, 2001) and as such contribute to the enhanced greenhouse effect, making the atmospheric emission from wildfires a major focus for global change scientists (Wooster, 2002). Recent work (e.g. Roberts et al., 2005) has shown that remote sensing, in combination with *in situ* and laboratory studies, provides a way to quantify not only the amount of biomass combusted during vegetation burning events, but also the amount of trace gases that enter the atmosphere as a result. This is because the total amount of energy produced by a fire is proportional to the amount of biomass consumed and the carbon dioxide volatilised (i.e. released as a gas during the process). By making measurements of the amount of energy produced during wildfires, remote sensing can, therefore, allow us to estimate the impact that wildfires and biomass burning in general have on the composition of the Earth's atmosphere.

Remote sensing the health of the Earth's coral reefs

Coral reefs are limestone structures secured to the ocean bed, and are composed of the skeletons of coral, carnivorous animals that live off zooplankton, and calcareous and non-calcareous algae. Although the three main reef morphologies – barrier, fringing and atoll – vary greatly in size, they all share a common characteristic: they form and thrive under a very limited range of environmental conditions. Coral reefs develop best in water that has a temperature of between 18 and 30 °C and normal levels of salinity, conditions which currently prevail in the tropical latitudes of between 30 °N and 25 °S. Coral reef communities also need light for photosynthesis, and as a result form only in shallow water that is relatively clear and free from significant amounts of suspended sediments (Souter and Linden, 2000).

Coral reefs are important for several reasons. They provide shelter for a huge number of species to live and reproduce, a wealth of biodiversity that has important implications not only for the marine ecosystem, but also for humans through the potential for pharmaceutical exploitation. Reefs also provide natural barriers to wave erosion for many low-lying island nations, the economies of which are often dependent on the opportunities that the reefs provide for the commercial fishing and tourism industries.

The specificity of the environmental conditions under which coral reefs develop and survive makes them sensitive indicators of environmental change. As a consequence, they are vulnerable to a range of factors, both natural, and anthropogenic, that may conspire to change their environment. Rising sea temperatures and elevated levels of atmospheric carbon dioxide, which limits the rate at which

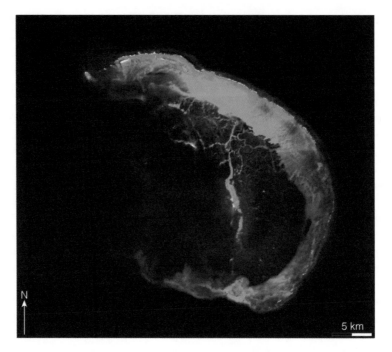

Figure 5.17   Landsat ETM+ true colour image of French Frigate Shoals. An open atoll consisting of a large crescent-shaped coral reef, French Frigate Shoals is part of the Northwestern Hawaiian Island chain, located about 800 km north-west of Honolulu.

reefs grow by reducing the rate of calcification (Kleypass et al., 1999), are forecast to threaten the health of the Earth's coral reefs over the coming decades (Mumby et al., 2002). Further stress will result from the combined effects of overfishing, over-tourism and pollution.

The importance of coral reefs and their value as indicators of environmental change has prompted scientists to invest a great deal of effort into compiling inventories of coral reef abundance, as well as detecting and monitoring changes in their health. Given the geographical extent of the area in which coral reefs exist, field surveys, using boats and divers, are both prohibitively expensive and time-consuming at anything other than the local scale. Satellite remote sensing, on the other hand, allows large areas to be surveyed quickly and in a cost-effective manner (although detailed large-scale mapping, and calibration of the remote sensing data, do still require a certain amount of *in situ* data collection; see Mumby et al., 1999). As the water that overlies the reef material is transparent only to radiation in the visible region of the electromagnetic spectrum, it is at these wavelengths that researchers have looked for changes in the aforementioned spectral signatures of coral that are diagnostic of changes in the health and composition of reef communities. Although the role of remote sensing in coral reef science remains to be fully developed, numerous studies have demonstrated the potential of remote sensing to characterise reef growth patterns and the influence that climate and weather exerts on them, ecosystem composition, and the interaction of coral reefs with sediment plumes, among others. Of particular interest is the phenomenon of coral reef 'bleaching'.

Bleaching describes the whitening of corals, due to the loss of symbiotic algae and/or their pigments (Brown, 1997). It results from an increase in either the temperature of the water or the amount of solar radiation that the coral receives, and as a result of the loss of algae, the coral changes from a brown or green colour to the white of the basic coral skeletons. If the bleaching is severe enough or sustained for a sufficiently long period of time, the coral can be killed. Several studies have demonstrated the ability and potential of remote sensing to identify this change in colour, or spectral reflectance, and as such can be used to monitor coral mortality (see Clark et al., 2000; Holden and Ledrew, 1999; Yamano and Tamura, 2004).

Although remote sensing has been demonstrated as an economical technique to map and monitor coral reefs, the success of the technique can be limited by factors including the depth and clarity of the water overlying the coral and overlaps (i.e. similarity) in the spectral response of materials on the ocean floor. However, the increased availability of high spatial resolution data (for example, the sub-metre spatial resolution images provided by commercially operated satellites such as IKONOS and QuickBird) allows more detail to be extracted. Increased spectral resolution, and in particular the availability of hyperspectral data (i.e. reflectance data in hundreds of contiguous wavebands), offer the possibility for improved remote sensing analysis of the extent and health of the Earth's coral reefs.

Summary

First, by now you will have realised that environmental technology is moving very quickly. By the time you read this, many advances will have been made beyond what is in this chapter and new technologies will be emerging. Secondly, you will have come to realise that technology is a double edged sword. It has an immense power to do good, in all kinds of spheres, as well as the environment. At the same time technology can be misused, either deliberately, to manipulate or conceal, or through incompletely understood, or inappropriate applications. One of the principal aims of this chapter has been to enable you to understand the issues sufficiently to make judgements about, and to evaluate the use of, environmental technology in a variety of situations.

Of course, you should also have learned something of the way in which the technologies function, sufficiently to understand what they do and how they do it, even though not becoming experts. You will have realised the way in which technologies are merging, alongside the development of new ones. You will also have become aware of the nature, and diverse sources, of geoinformation and its enormous and ever growing volume. Finally, it is hoped you will have acquired the perceptiveness and vision to see the potential for environmental geotechnology in your lives and the lives of others.

References

Andreae, G.P. and Merlet, P. (2001) 'Emissions of trace gases and aerosols from biomass burning', *Global Biogeochemical Cycles*, 15(4): 955–66.

Brown, B.E. (1997) 'Coral bleaching: causes and consequences', *Coral Reefs*, 16, Suppl: S129–38.

Burgess, E.W. (1925) 'The growth of the city. an introduction to a research project', in R.E. Park, E.W. Burgess and R.D. McKenzie (eds), *The City*. Chicago: University of Chicago Press.

Christaller, W. (1933) *Die zentralen Orte in Süddeutschland*. Jena: Gustav Fischer.

Clark, C.D., Mumby, P.J., Chisholm, J.R.M., Jaubert, J. and Andrefouet, S. (2000) 'Spectral discrimination of coral mortality states following a severe bleaching event', *International Journal of Remote Sensing*, 21: 2321–7.

Craig, W.J., Harris, T.M. and Weiner, D. (eds) (2002) *Community Participation and Geographic Information Systems*. London: Taylor & Francis.

Curry, M.R. (1995) 'GIS and the inevitability of ethical inconsistency', in J.Pickles (ed.), *Ground Truth: The Social Implications of Geographic Information Systems*. New York: The Guilford Press.

Dousset, B. and Gourmelon, F. (2003) 'Satellite multi-sensor analysis of urban surface temperatures and landcover', *ISPRS Journal of Photogrammetry and Remote Sensing*, 58: 43–54.

Eastman, J.R. (1997) *Idrisi for Windows User's Guide*. Worcester, MA: Clark University.

Ewing, R. (1997) 'Is Los Angeles-style sprawl desirable?', *Journal of the American Planning Association*, 63: 107–26.

Goudie, A. (2001) *The Nature of the Environment*, 4th edn. Oxford: Blackwell. p. 544.

Heywood, I., Cornelius, S. and Carver, S. (2006) *An Introduction to Geographical Information Systems*. Harlow: Pearson Education.

Holden, H. and Ledrew, E. (1999) 'Hyperspectral identification of coral reef features', *International Journal of Remote Sensing*, 20: 2545–63.

Imhoff, M.L, Lawrence, W.T., Stutzer, D.C. and Elvidge, C.D. (1997) 'A technique for using composite DMSP/OLS "city lights" satellite data to map urban area', *Remote Sensing of Environment*, 61: 361–70.

Kaufman, Y.J., Justice, C.O., Flynn, L.P., Kendall, J.D., Prins, E.M., Giglio, L., Ward, D.E., Menzel, P. and Setzer, A.W. (1998) 'Potential global fire monitoring from EOS-MODIS', *Journal of Geophysical Research*, 103: 32215–38.

Kleypass, J.A., Buddemeier, R.W., Archer, D., Gattuso, J-P., Langdon, C. and Opdyke, B.N. (1999) 'Geochemical consequences of increased atmospheric carbon dioxide on coral reefs', *Science*, 284: 118–20.

Longley, P.A., Goodchild, M.F., Maguire, D.J. and Rhind, D.W. (2005a) *Geographic Information Systems and Science*. Chichester: Wiley.

Longley, P.A., Goodchild, M.F., Maguire, D.J. and Rhind, D.W. (eds) (2005b) *Geographical Information Systems: Principles, Techniques, Management, and Applications*. Hoboken: Wiley.

McCall, M.K. (2004) 'Can participatory-GIS strengthen local level spatial planning? Suggestion for better practice', GISDECO 7th International Seminar: GIS for Developing Countries. Skudai, Johor, Malaysia.

Monmonier, M. (1996) *How to Lie with Maps*. Chicago: University of Chicago Press.

Mumby, P.J., Green, E.P., Edwards, A.J. and Clark, C.D. (1999) 'The cost-effectiveness of remote sensing for tropical coastal resource assessment and management', *Journal of Environmental Management*, 55: 157–66.

Mumby, P.J. and Edwards, A.J. (2002) 'Mapping marine environments with IKONOS imagery: enhanced spatial resolution can deliver greater thematic accuracy', *Remote Sensing of Environment*, 82: 248–57.

Oke, T.R. (1973) 'City size and the urban heat island', *Atmospheric Environment*, 7: 769–79.

Pedersen, D., Smith, V.E. and Adler, J. (1999) 'Sprawling, sprawling....', *Newsweek*, 23–7.

Pickles, J. (1995) *'Representations in an electronic age: geography, GIS and democracy'*, in J. Pickles (ed.), *Ground Truth: The Social Implications of Geographic Information Systems*. New York: The Guilford Press.

Pickles, J. (2005) *'Arguments, debates and dialogues: the GIS–social theory debate and the concern for alternatives'*, in P.A. Longley, M.F. Goodchild, D.J. Maguire and D.W. Rhind (eds), *Geographical Information Systems: Principles, Techniques, Management, and Applications*. Hoboken, NJ: John Wiley & Sons.

Roberts, D.A., Dennison, P.E., Gardener, M.E., Hetzel, Y., Ustin, S.L. and Lee, C.T. (2003) 'Evaluation of the potential of Hyperion for fire danger assessment by comparison to the Airborne Visible/Infrared Imaging Spectrometer', *IEEE Transactions on Geoscience and Remote Sensing*, 41: 1297–1310.

Roberts, G., Wooster, M., Perry, G., Drake, N., Rebelo, L-M. and Dipotso, F. (2005) 'Retrieval of biomass combustion rates and totals from fire radiative power observations: application to southern Africa using geostationary data', *Journal of Geophysical Research, sub judice*.

Rosenfeld, A.H., Akbari, H., Bretz, S., Fishman, B.L., Kurn, D.M., Sailor, D. and Taha, H. (1995) 'Mitigation of urban heat islands: materials, utility programs, updates', *Energy and Buildings*, 22: 255–65.

Schuurman, N. (2004) *GIS: A Short Introduction*. Oxford: Blackwell.

Smith, K. (2001) *Environmental Hazards: Assessing Risk and Reducing Disaster*. London: Routledge. p. 392.

Souter, D.W. and Linden, O. (2000) 'The health and future of coral reef systems', *Ocean and Coastal Management*, 43: 657–88.

Sparkes, A. and Kidner, D. (2006) 'A GIS for the environmental impact of wind farms', http://gis.esri.com/library/userconf/europroc96/PAPERS/PN26/PN26F.HTM

Streutker, D. R. (2003) 'Satellite-measured growth of the urban heat island of Houston, Texas', *Remote Sensing of Environment*, 85: 282–9.

Sutherland, E.R., Make, B.J., Vedal, S., Zhang, L., Dutton, S.J., Murphy, J.R. and Silkoff, P.E. (2005) 'Wildfire smoke and respiratory symptoms in patients with chronic obstructive pulmonary disease', *The Journal of Allergy and Clinical Immunology*, 115: 420–2.

Sutton, P.C. (2003) 'A scale-adjusted measure of "urban sprawl" using night-time satellite imagery', *Remote Sensing of Environment*, 86: 353–69.

Voogt, J.A. and Oke, T.R. (2003) 'Thermal remote sensing of urban climates', *Remote Sensing of Environment*, 86: 370–84.

Wilson, E.H., Hurd, J.D., Civco, D.L., Prisloe, M.P. and Arnold, C. (2003) 'Development of a geospatial model to quantify, describe and map urban growth', *Remote Sensing of Environment*, 86: 275–85.

Wooster, M.J. (2002) 'Small-scale experimental testing of fire radiative energy for quantifying mass combusted in natural vegetation fires', *Geophysical Research Letters*, 29, doi: 10.1029/2002GL015487.

Yamano, H. and Tamura, M. (2004) 'Detection limits of coral reef bleaching by satellite remote sensing: simulation and data analysis', *Remote Sensing of Environment*, 90: 86–103.

HELPFUL WEBSITES

http://www.ccg.leeds.ac.uk/groups/democracy/work.html
http://www.ccg.leeds.ac.uk/teaching/wilderness/wildbritain.html
http://www.groundwork-tv.org.uk/downloads/Groundwork%20Review%202004.pdf
http://ppgis.iapad.org/
http://rangeview.arizona.edu/About/index.asp

SECTION 2
CASE STUDIES

6 Food

Susan Buckingham and Phil Collins

Learning outcomes

Knowledge and understanding of:

o   contemporary patterns of food consumption habits in the Global
    North, and the impact this has on societies and environments across
    the world
o   the impacts of industrial food production on the environment.

Critical awareness and evaluation of:

o   the environmental impacts of transporting food
o   local and global inequalities in food security
o   attempts to redress some of the social and environmental problems
    identified with industrial food production.

Introduction

In the summer of 2003, one of this chapter's authors was given a bottle of 'Fiji natural artesian water' in Colorado. This water was reputedly drawn from an aquifer 'deep beneath volcanic highlands and pristine tropical forests on the main island of Viti Levu in Fiji'. The distance of Fiji from any landmass was marketed as a virtue: 'Separated by over 1500 miles of the open Pacific from the nearest continent, this virgin ecosystem protects one of the purest waters in the world.' To reach Colorado, this water (also bottled in Fiji) had, in fact, to travel 6255 miles, or 10,066 kilometres, contributing to greenhouse gases on the way. Two years later, this brand of bottled water was on sale in London, UK – a distance of 10,099 miles, or 16,253 kilometres from Nanda in Fiji.

This vignette raises a number of issues that this chapter will discuss, including food consumption habits in the West and the environmental impacts of long-haul food transportation. While water is not generally classified as food, and easily warrants a chapter to itself (one that this book does not have the space to grant), it is (with oxygen) the most basic element that we all consume, and UNESCO estimates that we each need between 20 and 50 litres of water, free from contaminants, each day for our basic needs. Nevertheless, a child born in the developed world is likely to consume 30 to 50 times as much as a child born in the developing world. The example of 'pure' water from Fiji can be used to illustrate this inequity further. Although Fiji has a relatively small population (0.8 million), with a modest annual population growth of 1.2%, 23% of this population were without access to safe water in 1997 (Millstone and Lang, 2003). According to the UNDP Human Development Report (2003), in 2002, 53.1% of Fijians had no sustainable access to an improved water source, having therefore to rely on unimproved sources, including water vendors, bottled water, tanker trucks and unprotected wells and springs. Table 6.1 shows a number of indicators for Fiji, which is ranked 92 in the UNDP Human Development Index (HDI). It is salutary (and somewhat ironic, given the story of the water transfer) that the USA and UK, ranked 7 and 13 respectively on the HDI, and importers of 'Fiji natural artesian water', register no population without sustainable access to an improved water source. However, as the later section on food inequalities will show, there are significant differences within countries like the UK and USA when it comes to food consumption.

Table 6.1 also reviews a number of countries spanning the range of 'human development' as defined by the UNDP. They have been selected for their position in the ranking, their range of social, political and development/historical experience, and the relevance of issues dealt with in this chapter to these countries' experiences. This table will be referred to throughout the chapter, and readers will be invited to think about the relationships between the indicators shown.

This chapter addresses changing forms of food consumption, production and transportation which provide the context for discussing the environmental impacts of these practices and – a central theme which runs through this book – the ways in which these impacts are experienced unevenly by different communities. The role of international organisations in these processes is examined, and we conclude by exploring the potential of alternative food production strategies and networks to overcome some of the problems identified.

Before proceeding, it is important to remember the underlying economic and political structures in and through which patterns of food production, transportation and consumption are constructed, and which consequently determine broad patterns of human and environmental health and well-being. There are clear patterns in the relationship between countries in the North and the South, between large-scale food corporations and small producers, and between the rich and poor within any one country. These relationships both structure the food supply chain and affect the environment as a result.

Contemporary Western food consumption habits

Over the last few decades, food has not only become more varied, but cooking it has become a hobby for many men and women. As this chapter was being written, the *Observer*'s list of five best-selling manuals for 29 January 2006 included

Table 6.1 Selected countries' experience of indicators relevant to food

Country (HDI rank[1])	Water[2] (% rural/urban)	Daily Calories pp[3]	Under nutrition[4]	Over nutrition[5]	Agro chemicals (pesticides/fertilisers[6])
Norway (1)	100	3357	–	3.9	941/218
The Netherlands (5)	100	3284	–	3.9	11,842/821
The USA (7)	100	3699	1	4.8	1599/151
The UK (13)	100	3276	–	3.5	4745/343
Italy (21)	ND	3507	–	5.1	19,288/227
Israel (22)	ND	3278	–	4.4	–/360
South Korea (30)	100	3313	2	3.1	13,829/693
Hungary (38)	98/100	3313	2	4.5	2863/83
Costa Rica (42)	92/99	2649	2	5.4	18,726/322
Cuba (52)	77/95	2480	9	4.1	–/52
Mexico (55)	69/95	3097	14	3.3	–/54
Belize (67)	82/100	2907	6	2.0	17,804/50
Thailand (74)	81/95	2360	19	5.1	1116/75
Oman (79)	30/41	–	23	5.8	24,125/122
Fiji (81)	51/43	2865	8	2.5	2333/66
South Africa (111)	73/99	2990	9	2.7	57/51
India (127)	79/95	2496	53	0.4	436/89
Uganda (147)	47/80	2085	26	0.5	17/0
Malawi (162)	44/95	2043	30	0.4	–/31
Niger (174)	56/70	2097	43		–/2

Notes:
[1] UNDP, 2003
[2] Percentage of population with sustainable access to safe water in 2000, Millstone and Lang, 2003
[3] Average daily calorie supply per person in 1997 from World Resources 2000–2001 (Table AF3 in Millstone and Lang, 2003)
[4] Percentage of children underweight in 1990–1997 (Millstone and Lang, 2003)
[5] Percentage of people with diabetes in 2000 (WHO Statistical Information Service, 2001 in Millstone and Lang, 2003)
[6] kg used per hectare of cropland, World Resources 2000–2001 (Table AF2 in Millstone and Lang, 2003)
– indicates no data available

three cookery books (and one gardening book). The taste for more exotic food, stimulated variously by the media industry, food company marketing and increasing numbers of holidays taken overseas, creates a range of environmental problems, from pollution generated by the transportation of foodstuffs to the intensive use of land and artificial inputs used to grow crops for export.

Apart from the two world wars, which imposed severe dietary restrictions across most of Europe (although these restrictions in the UK, which applied to butter, sugar and meat and required all available land to be turned over to growing vegetables and fruit, led to a healthier diet than many families are currently used to), families in early 20th century Britain ate well on mostly local produce. Our own parents, now in their 80s and 90s, remember the meals of their childhood: vegetables grown in local allotments, fish and seafood from Cardiff Bay or the North Sea, cakes baked in the local baker's oven and a few imported luxuries such as bananas and pineapples. Cooking was a skill passed from mother to daughter and, in the home at least, was considered a job for women.

The difference now is that cookery is marketed as a middle-class leisure activity for women and men which contributes significantly to the economy in terms of books sold, TV programmes watched and produce bought. The irony of this situation is that in countries such as the UK and USA, most families are increasingly 'time poor' and rely more and more on prepared food, rich in preservatives such as salt and sodium, and heavily packaged. The rest of this section will focus on two consequences of these developing food consumption habits: the health implications of a Western diet characterised by high meat content and high levels of additives, and the environmental impact of the packaging used to transport and market food (which Chapter 7 on waste develops in more detail).

Health impacts

Table 6.1 has indicated a close correspondence between a higher human development ranking and diabetes, a frequent consequence of obesity. Millstone and Lang (2003) also present data which show that 21 per cent of women and 17 per cent of men are considered obese in the UK, while in the USA, the figures are 25 per cent of women and 20 per cent of men. An American anti-hunger campaigning group, Project Bread, has assembled evidence suggesting that obesity is often an environmentally (as opposed to genetically) caused condition related to increased consumption of fast food and carbonated drinks, insufficient consumption of fruit and vegetables and a more sedentary lifestyle with less physical activity (Project Bread, 2004). This evidence is also used to demonstrate the link between poverty and obesity, which this chapter will later examine in more detail with regard to inequality.

Children habituated to a fast-food diet are increasingly vulnerable to obesity and its corollaries: type 2 diabetes and cardiovascular disease. The UK Food Standards Agency has warned against feeding children the 'big five' of bad food: sugared breakfast cereals, soft drinks, confectionery, savoury snacks and fast foods (FSA, 2003). A belated acknowledgement of the link between advertising and the increased consumption of these foods was made in November 2006 in the

UK, when the government decided to ban the advertisement of fast foods during children's television broadcasting.

A report by the FAO/World Bank in 1998 referred to 'the nutrition transition' in which countries, as they get richer, increase their consumption of meats and fats, relative to carbohydrates. Urbanisation also appears to trigger a change in diet, with more sugar consumed (Popkin, 1999). As well as the diseases referred to above, this nutrition transition is also leading to an increased incidence in cancers. For example, the five most widespread cancers worldwide (in order of incidences: lung, stomach, breast, colon/rectum, mouth/pharynx) are all associated with low vegetable and fruit intake; breast cancer is also linked to obesity and colon/rectum cancer with increased meat intake. The nutrition transition has environmental impacts also as the land needed to feed and accommodate animals is much greater than for growing cereals.

Arguably, these health risks are within people's control (taking into account the power of advertising and marketing by the food industry, and household income), and yet much greater public concern appears to be generated by the increasing number of food scares in recent decades. This relates to societies' perception of risk, a concept reviewed in the opening chapter of this book. Suzanne Freidberg calls the public reaction to food scares summarised in Table 6.2 'acute collective anxiety', which has little to do with 'proven danger: none of the diseases implicated in Europe's late twentieth-century food scares had yet sickened or killed anywhere near as many people as, say, obesity or smoking' (Freidberg, 2004: 5–6).

Food packaging

According to a report for *The Observer Magazine*, the UK throws away 4.6m tonnes of packaging waste per year, adding around £480 a year to its average food bill (Siegle, 2006). The packaging industry, which is self-regulating, defends its practice by arguing that packaging protects 10 times its own weight of goods and that around 63% of packaging in the UK is now recycled. The UK packaging industry, in attempts to prevent legislation which it believes would not be in its favour, has negotiated a 'producer responsibility' to reduce waste. The Courtauld Commitment (see Box 6.1) is an example of this producer responsibility, and the quote from the government minister then responsible indicates the complexity of the relationship between the industry and the government. It is worth dwelling on this box to think through what this relationship might involve. Despite the existence of the producer responsibility and the Courtauld Commitment, increasingly sophisticated packaging (which, for example, demands 14.4 grams of oil to manufacture a Heinz tomato ketchup 'stay clean' cap, compared to the 3.8 grams needed for a 'normal' lid) competes with an ongoing campaign to reduce the amount of packaging involved in the food industry, led by environmental, women's and consumer campaigning groups, such as the Women's Environmental Network and the National Federation of Women's Institutes. At the end of the food consumption chain, consumers worldwide use 750 billion carrier bags each year which results in problems both at the production end (the use of oil to manufacture these) and in their disposal. Many of these bags end up damaging

Table 6.2   Summary of recent food scares in the UK

Year	Disease	Vector	Effects on humans and environments
1988	Salmonella	Chickens, eggs	Poisoning
1988	Listeria	Pate, soft cheeses	Poisoning
1986	BSE	Cows	CJD lethal disease (from 1996)
2001	Foot and mouth	Hoofed animals	No health effect, but widespread culling of animals and countryside off-limits

wildlife when they are discarded and find their way into rivers and seas, and Greenpeace has drawn attention in 2006 to the phenomenon they have called 'the trash vortex' in the South Pacific, in which a slowly circulating current of plastics waste in the North Pacific Gyre between Hawaii and the Californian coast is causing problems for marine wildlife (Greenpeace, 2006). Ireland's now famous levy on plastic bags has sought to reduce this problem and is said to have reduced plastic bag use in Ireland by 90 per cent (Siegle, 2006).

Box 6.1   The packaging industry's response to waste reduction

Courtauld Commitment:

In July 2005, 13 major UK retailers (12 of which sell food and which include all the big supermarket chains) joined forces with WRAP (Waste Reduction and Packaging), in pledging to tackle packaging and food waste to, among other commitments, identify ways to tackle the problem of food waste. Ideas ranged from reducing layers of packaging to light-weighting some packaging materials and initiating research into why consumers throw away an average £400-worth of food every year.

Elliot Morley, then Minister for Climate Change and Environment in the UK, offered his support for the commitment:

'Our food and drink industry is a major employer and a major part of our economy, but it is also a major source of waste – accounting for about ten per cent of all industrial and commercial waste, notably packaging.

'I am encouraged by the response of the major retailers, who are taking up the challenge to shrink the amount of packaging and food waste produced both by the industry and by consumers.

'Much of the waste we generate in the home starts as food and packaging we buy in the supermarkets, so the industry, from retailers all the way up the supply chain, is in a prime position to influence consumer behaviour. They are already making significant environmental improvements which are good for business, and this commitment is a positive step down the road to tackling the waste problem.'

Source: Wrap 2005

The industries referred to above are part of a global food production system which is the focus of the next section.

Food production processes

In order to supply the West's increasing demand for exotic and out-of-season food, and also hoping to promote this demand, a global production market has emerged, mainly controlled by a small number of large conglomerates which command wide market share and control of most functions from planting to selling food. Producers worldwide are under pressure to deliver food which meets these firms' particular specifications, and which usually involves the use of pesticides and fertilisers (themselves mostly oil derivatives), and overexploitation of natural resources which creates negative effects on the wider environment. These industrial inputs are one characteristic of one of three 'food paradigms' (see Box 6.2) identified by Tim Lang and Michael Heasman in their book *Food Wars*, published in 2004. This is the 'productivist paradigm' which dominated most of the middle and late 20th century and focused on food producers and chemical inputs. This paradigm is also characterised by production concentrated in relatively few hands, mass distribution of foodstuffs, and monocultures. The other two paradigms are the 'Life Sciences Integrated Paradigm', characterised by biological rather than chemical inputs (as the discussion on genetic modification will demonstrate shortly), and the 'Ecological Paradigm' which recognises diverse natural communities, the importance of local skills and management, the nutrient cycle and natural control mechanisms. Present-day Cuba is a good example of the ecological food paradigm and will be discussed at the end of this chapter. Before addressing the activities of food manufacturers and retailers, this section will first consider two strategies which have been promoted, primarily by politicians and multinational companies, to generate industrial-scale food production which, they argue, is needed to feed the world's growing population. Also, as a later section will show, these strategies carry their own environmental problems.

Box 6.2   Food paradigms

Lang and Heasman use this term to 'indicate a set of shared understandings, common rules and ways of conceiving problems and solutions about food'. Drawing on Thomas Kuhn's work on scientific paradigms, which represent a collective agreement around an approach to knowledge which prevails at any given time, they summarise a paradigm, for their purposes, as a 'fundamental set of framing assumptions that shape the way a body of knowledge is thought of'. (Lang and Heasman, 2004: 17)

The Green Revolution and genetic modification

The 'Green Revolution' refers to an industrialised form of farming which relies on heavy use of artificial inputs. It is characterised by monocultures, which, since they are more predisposed to pests and disease, require the application of artificially produced fertilisers and pesticides. According to Vandana Shiva, an Indian physicist who is also a renowned environmental and human rights campaigner, the Green Revolution was so-called as it was a deliberate strategy supported by Western governments, particularly the USA, to combat socialism (the 'Red Revolution'), through agriculture (Shiva, 2002). However, from Table 6.1, which has presented a range of food-related data, it becomes clear that the amount of artificial pesticides and fertilisers applied bears little relationship to food availability or the general health of populations. For example, the country with the highest rate of pesticide application worldwide (Oman) also has an exceptionally high rate of under nutrition, measured by the percentage of children recorded as underweight.

Western governments and multinational companies justified their interventions by the alleged need to grow food on an industrial scale to ensure sufficient supplies for the world's fast growing population. Such a claim is undermined by Amartya Sen's Nobel prize winning work on famine, which argues that food shortages are rarely linked to absolute lack of food, but rather uneven food distribution, whereby the poor are unable to pay the inflated price of food in certain situations (Drèze and Sen, 1989). Some of the thinking which underlay this Green Revolution was that prosperity would be brought to agriculture which would deflect farmers' demands for land reform, or support for communist governments, particularly in countries such as India. However, far from creating a satisfied community of farmers, the consequences of creating a reliance on artificial inputs creates a high degree of vulnerability to price fluctuations. Shiva reports that the huge rise in pesticide use in India (up by 6000 per cent over a period of 20 years) is leading to increases in debt for Indian farmers, which she claims accounted for around 20,000 suicides over a two- to three-year period (Shiva, 2002). Multinational companies stood to gain substantially from the Green Revolution, including companies such as Union Carbide (now owned by Dow Chemicals), whose pesticide manufacturing plant in Bhopal, India exploded in 1984, causing a high number of deaths and injuries, and substantial environmental problems, as Box 6.3 illustrates. Further environmental problems linked to the Green Revolution are explored later in the chapter.

Box 6.3   The legacy of the Union Carbide plant explosion in Bhopal, India

'On the night of Dec. 2nd and 3rd, 1984, the Union Carbide pesticide plant in Bhopal, India, began leaking 27 tons of the deadly gas *methyl isocyanate*, which spread throughout the city of Bhopal. Half a million people were exposed to the gas and 20,000 have died to date as a result of their exposure. More than 120,000 people still suffer from ailments caused by the accident and the subsequent pollution at the plant site, including blindness, extreme difficulty in breathing, and gynaecological disorders. The site has never been properly cleaned up and it continues to poison the residents of Bhopal. In 1999, local groundwater and wellwater testing near the site of the accident revealed mercury at levels between 20,000 and 6 million times those expected. Cancer and brain-damage-and birth-defect-causing chemicals were also found in the water.

'The Union Carbide factory in Bhopal seemed doomed almost from the start. The company built the pesticide factory there in the 1970s, thinking that India represented a huge untapped market for its pest control products. However, sales never met the company's expectations; Indian farmers, struggling to cope with droughts and floods, didn't have the money to buy Union Carbide's pesticides. The plant, which never reached its full capacity, proved to be a losing venture and ceased active production in the early 1980s . . . In December 1999, Greenpeace reported that soil and water in and around the plant were contaminated by organochlorines and heavy metals. A February 2002 study found mercury, lead and organochlorines in the breast milk of women living near the plant. The children of gas-affected women are subject to a frightening array of debilitating illnesses, including retardation, gruesome birth defects, and reproductive disorders. It wasn't until 1989 that Union Carbide, in a partial settlement with the Indian government, agreed to pay out some $470 million in compensation. The victims weren't consulted in the settlement discussions, and many felt cheated by their compensation – $300–$500 – or about five years' worth of medical expenses. Today, those who were awarded compensation are hardly better off than those who weren't . . . .

'Union Carbide also remains liable for the environmental devastation its operations have caused. Environmental damages were never addressed in the 1989 settlement, and the contamination that Union Carbide left behind continues to spread. These liabilities became the property of the Dow Corporation, following its 2001 purchase of Union Carbide. The deal was completed much to the chagrin of a number of Dow stockholders, who filed suit in a desperate attempt to stop it . . . . Dow was quick to pay off an outstanding claim against Union Carbide soon after it acquired the company, setting aside $2.2 billion to pay off former Union Carbide asbestos workers in Texas. However, Dow has consistently and stringently maintained that it isn't liable for the Bhopal accident.'

Source: Lapierre and Moro (2002)

While the Green Revolution is an example of Lang and Heasman's 'productivist paradigm', genetic modification of food products illustrates their 'life sciences integrated paradigm', in which chemical inputs are replaced by biological inputs. Genetic modification (GM) is a process which takes genes from unrelated species and inserts them into another to create a transformed plant or animal, the purpose of which is to introduce a particular characteristic to the host plant (such as disease resistance). As with the Green Revolution, Western governments and multinational companies justify this development on the basis that the world will soon be facing food shortages. However, there are considerable concerns that GM increases the dependency of farmers on the biotechnology industry, for example, through its property rights over the seeds which farmers need to buy annually from the companies who produce the genetically modified seeds (which are programmed not to reproduce). GM also carries a number of environmental risks linked to introducing modified plant species into existing habitats, which are currently poorly understood.

Fast progress has been made such that more than 50 million hectares world-wide are now under GM crops, from zero in the mid-1990s, with 68 per cent of these in the USA. Globally, 40 per cent of all soy produced is genetically modi-fied, 7 per cent of maize, 20 per cent of cotton and 11 per cent of oilseed rape (Lang and Heasman, 2004). However, uptake has been much more cautious in Europe, partly because of significant popular concern and an effective boycott of GM food products (see Chapter 3 for a discussion on the influence of environ-mental campaigning groups).

Both the Green Revolution and genetic modification have been largely respon-sible for the increasing industrialisation of food production, with its demand for artificial inputs and pressure on water supplies, the latter of which is a factor in the increase in the number of large dam projects, such as the Narmada dam pro-ject presented in Chapter 3. They are also responsible for the clearance of land, including deforestation, for intensive agriculture and the shift from subsistence to commercial agriculture. Often this drives subsistence agriculture into more marginal land, reducing its productivity, and creates longer distances between subsistence farming and water sources. There are also impacts on the wider environment, either by removal of resources, or by the addition of waste gener-ated during production, harvesting and processing which will be considered later.

The food manufacturing and retailing industry

Although the food industry cannot compete with industrial sectors, such as aero-space, automotives, chemicals or defence in terms of turnover, there is sufficient consolidation around the largest companies to give them significant influence. Table 6.3 lists the world's top twenty food groups.

It is an interesting exercise to consider the environmental performance of these companies, using the UK-based *Ethical Consumer* magazine's database 'Ethiscore', which can be seen from Table 6.3 to have scored a number of them. Pepsi-Cola is an interesting case in point, and should be of particular interest to students on at least two counts: firstly, the volume consumed on campuses in the UK and USA suggests that this is a highly popular drink among students. Secondly, and a likely contributor to this, in the USA, many schools have struck deals with the Pepsi-Cola Company, whereby Pepsi-Cola have exclusive vending rights in exchange for an undisclosed sum of money (Klein, 2001: 91). Referring back to an earlier sec-tion which demonstrated the negative health impacts of carbonated drinks, this raises an even more significant issue concerning the relationship between schools and multinational companies.

Further evidence of the consolidation of international food production and trade is available by examining the number of companies which account for the majority of the food industry. For example, Barbara Adams reports that, world-wide, only six companies dominate the grain trade, while fifteen companies account for eighty per cent of all world food trade (Adams, 1998). Like food man-ufacturers and distributors, food retailers are becoming increasingly consolidated

Table 6.3 The world's top twenty food groups in 2000

Rank	Company	Total Sales ($ millions)	Food Sales ($ millions)	Ethical Score[1]
1	Nestlé	47,489	44,640	0
2	Philip Morris Co. Inc.	80,356	26,532	NS
3	ConAgra Foods Inc.	24,535	25,535	NS
4	Cargill	47,602	22,500	NS
5	Unilever Bestfoods	41,403	20,712	3
6	The Coca-Cola Company	20,458	20,458	3
7	PepsiCo	20,438	20,438	1.5
8	Archer Daniels Midland Company	20,051	20,051	4
9	IBP Inc.	16,950	16,950	NS
10	Arla Foods	15,824	15,824	12.5
11	Diageo	16,938	15,584	7.5
12	Mars	15,300	15,300	3.5
13	Anheuser-Busch	12,262	12,262	4.5
14	Danone	12,576	12,123	NS
15	Kinn Brewery Company	12,826	11,928	NS
16	HJ Heinz Company	9430	9430	NS
17	Asahi Breweries	11,351	9194	NS
18	Suntory	10,485	8823	NS
19	Snow Brand Milk Products	9256	8516	NS
20	Sara Lee Corporation	17,511	7705	6

Sources: Ethical Consumer Magazine, 2006; Lang and Heasman, 2004

Notes:

[1]Ethiscore allocate a score from 0 (worst performance) to 20 (best performance) to a range of companies, according to a set of criteria reflecting environmental and human rights behaviour. (NS indicates no score available)

and, with that, they acquire considerable power. The US supermarket chain, Wal-Mart, for example, is the largest global food retailer with a turnover of $180,787 million, operating in ten countries in 2002. Tesco, the largest UK food retailer is the world's tenth largest food retailer with a turnover of $31,812 milion and operating in nine countries. In the UK, the four largest supermarkets – Tesco, Sainsbury, Asda (a subsidiary of Wal-Mart) and Morrisons-Safeway – accounted for 73.3% of spending in the supermarket sector in 2006, with Tesco's share (31.3%) twice as large as its closest competitor (The *Guardian*, 2006). The power

that food retailers exert can be seen from the practices they have adopted to
squeeze the competition, as Box 6.4 shows by quoting at some length from Klein's
powerful polemic against the power of the multinationals.

Box 6.4   Securing market share: the Wal-Mart model

'The recipe that has made Wal-Mart the largest retailer in the world, hauling in $137 billion
in sales in 1998, is straightforward enough. First, build stores two and three times the size
of your closest competitors. Next, pile your shelves with products purchased in such great
volume that the suppliers are forced to give you a substantially lower price than they would
otherwise. Then cut your in-store prices so low that no small retailer can begin to compete
with your "everyday low prices . . ."

'The argument against Wal-Mart's retail style – by now almost as familiar as Wal-Mart
itself – holds that bargain prices lure shoppers to the suburbs, sucking community life
and small businesses out of the town centers. Small businesses can't compete – in fact,
many of Wal-Mart's competitors claim that they pay more for their goods wholesale than
Wal-Mart charges retail . . . .

'Wal-Mart and similar retailers are all known in the retail industry as "category killers"
because they enter a category with so much buying power that they almost instantly kill
the smaller competitors.'

Naomi Klein (2001: 135)

While the practices illustrated in Box 6.4 could be said to make cheap food available to a much wider population in the West, it does so at some cost to the producers (as Lang and Heasman point out in *Food Wars*), and to the poorest
communities who live in what have popularly come to be known as 'food deserts'.
Before discussing this, however, the next section turns to some of the environmental implications resulting from the practices identified so far in food production and consumption.

Environmental impacts

Pressure on water supplies

Since prehistoric times, people have been using and managing their water
resources for irrigation, fishing, drinking, cooking and, ultimately, waste disposal.
These early examples of water utilisation undoubtedly had an impact upon the
local environment, but they were still largely dependent on natural processes,
particularly rainfall. As a result, even powerful states were vulnerable to prolonged droughts.

The onset of industrialisation saw increasing independence from climatic shifts, or
at least ones lasting just a few years. Industrial agriculture in the late 20th century

and early 21st century requires three to five times more water for producing the same amount of food, compared to non-industrial agriculture. This is partly due to the use of organic manure in the latter, which is a much more effective conserver of water (Shiva, 2002). Improvements in pumping technology meant water could not only be transported greater distances across the land surface, but could also be extracted cheaply from deep below the ground. Water extraction enabled arable crops to be cultivated in areas that would earlier have been considered unsuitable. Even apparently inhospitable semi-arid environments could be used for food production.

Reliance on groundwater supplies, however, can have negative effects. Extraction at rates faster than the natural recharge leads to what is known as 'draw down'. This increase in extraction depth to saturated rocks (the aquifer) initially occurs in the vicinity of the borehole but can eventually spread out to affect the whole aquifer. The environmental impacts of this subsurface change can be considerable, including dieback of vegetation, soil degradation and subsidence, a situation particularly severe in Mexico City as the final chapter demonstrates. Solutions to this include slowing, or stopping, extraction to allow the aquifer to recover while other resources, including surface reservoirs, are used. Many countries are so heavily reliant on groundwater that this type of measure has to be accompanied by attempts to reduce consumption, such as the periodic, and controversial, hosepipe bans that affect parts of the UK.

Land clearance, land use change and biodiversity

Harvesting of food inevitably leads to ecological changes. Some of the best evidence for early occupation in north-west Europe by hunter-gatherers, for example, is the decline in tree pollen frequency, often associated with an increase of charcoal, suggesting more frequent fires. The onset of more formal agriculture leads to a much more fundamental, even catastrophic, ecological change as previously dominant vegetation communities are replaced by domesticated species which are systematically harvested. The immediate effect of clearance is often the removal of 'keystone' species that define an ecosystem, such as forest trees, and an overall reduction in biodiversity.

Pests and diseases tend to have an inverse relationship with plant diversity. Plant diversity creates an hospitable environment not only for pests, but also for their predators. Organic manuring (as opposed to artificial fertilising) also enriches the diversity of the soil, which itself plays a part in the control of pests. Destroying this diversity means disrupting the habitat for these predators, introducing the need for artificial pest control.

Recovery from land clearance through natural succession is prevented if agriculture takes place. Agriculture can be defined as the systematic exploitation of organic resources (i.e. food production), aided by technology, an important consequence of which is that a large part of the organic material produced is removed. This means that the ecological, or 'trophic' cycle of production, consumption and decomposition is disrupted. As a consequence, there may be a net loss of organic material, and the associated nutrients, from an area.

A further impact comes from the physical disruption of the soil by, for example, ploughing. Even apparently low technology disruption of soil structure can cause its degradation, as the increased contact of soil particles with the air leads to increased decomposition of organic matter, an important binding agent. Bare soil is also much more prone to erosion than where a vegetation layer is present. The onset of agriculture during the Neolithic and Bronze Age in southern Britain was marked by a massive increase in soil erosion and the Dust Bowl in North America in the 1930s was a spectacular example. Similar effects can be seen today in areas of cleared tropical rainforest.

Although the rate of erosion does tend to decline over time, there is no evidence that it ever returns to pre-clearance levels while farming continues. Erosional rills and gullies are common features in the fields of southern Britain, even after 5500 years of farming. Indeed, the increased intensification and mechanisation of farming in many areas during the 20th century led to an increase in soil erosion, partly because soils were much more intensely disrupted, for example, by deep ploughing. The new technology meant that marginal land could be developed, particularly during more favourable climatic periods.

Changes to the aquatic environment

Water bodies, whether fresh or marine, have always provided a food resource to communities. As with any intervention in an ecological system, this food extraction produces physical, chemical and biological impacts. Although the potential extent of these impacts is only just beginning to be understood, there has been a long history of attempting to mitigate human impacts on aquatic resources. The International Council for the Exploration of the Sea, for example, has provided a focus for marine research in the North Atlantic region since being established in 1902 (ICES, 2006), producing advice on a range of issues to a variety of stakeholder groups. Despite this, exploitation of the marine environment continues to have serious detrimental effects. North Sea cod stocks fell by 76% between 1963–2001 (National Statistics, 2006), largely as a result of overfishing, and recent research suggests recovery may be hampered by fishing-induced demographic changes within fish populations (Levin et al., 2005). Fishing quotas are a commonly proposed solution, though their implementation is controversial because of their inevitable economic and social impacts. Quotas have also led to increased discard – the dumping of undersized and non-quota fish overboard. Although some bird populations have benefited from this discard, fishing can have a detrimental impact upon sea-birds and other marine organisms, by removing their food source and by inadvertent trapping in fishing gear (Carranza et al., 2006).

Use of chemicals

Many pesticides and fertilisers are toxic chemicals which, in their use, carry health risks to humans and other species, reduce biodiversity, interfere with natural pest control, and pollute water. In addition, as the explosion of methyl-isocyanide at Bhopal in December 1984 very graphically demonstrated and continues to demonstrate (see Box 6.3), their manufacture and storage also carry significant environmental and health risks.

Millstone and Lang (2003) report that 20,000 agricultural workers are killed each year as a direct result of pesticide poisoning. While DDT was banned for sale in the West in the 1970s, it has remained in use elsewhere (partly to combat malaria, as well as as an agricultural pesticide). This has had severe health implications, as Rachel Carson argued in her landmark book *Silent Spring* (Carson, 1962), discussed in Chapter 3. Because the chemical load of DDT is inadvertently passed by mothers to their offspring through the placenta and through breast milk, these children are exposed to an increased number of diseases, including endometriosis and breast cancer in female children and reduced fertility. Other chemicals also cause concern: Sascha Gabizon (1998) has reported high concentrations of chemicals from pesticides in the Aral Sea region which coincide with higher than average rates of blue baby syndrome, maternal and infant mortality. Chemicals used in artificial pesticides are all implicated in female infertility and contaminated breast milk, and one of them, benzene, is also a well established carcinogen with no safe lower limit of ingestion.

The ways in which chemicals impact on different human bodies is a good example of how environmental problems are gendered in the way in which they are experienced, as will be demonstrated later, with respect to differences in food availability.

Transporting food

Food has long been transported across continents, and the diet we take for granted in the West today includes many staples not originally native to the countries in which they are now widespread. Given the expense of this transportation, its distance and the time it would take, it made sense to trade only in foodstuffs which were not available at home. Indeed, many traders were commissioned to bring home seeds, often secretly, which could then be planted locally, obviating the need for costly and time-consuming haulage. Compare this with an example of contemporary food trading that involves the Dutch importing ginger biscuits from the USA, and the Americans who were importing ginger biscuits from the Netherlands, prompting an exasperated Herman Daly to ask 'why they didn't just trade recipes?!' (Lucas, 2001, speech). If this sounds like an artfully constructed anecdote, consider the following: in 1997 over seven million tonnes of milk and cream were moved between European Union member states. In 1997, 126 million litres of liquid milk were imported into the UK *at the same time as* 270 million litres of liquid milk were exported out of the UK (Lucas, 2001). While there may be persuasive arguments to import exotic foodstuffs (and some of these arguments will be discussed later), the importing of indigenous foodstuffs is harder to defend. Another of Lucas's examples of wasteful food swapping is apples, in which she questions the logic of destroying 60 per cent of UK orchards since 1970, while importing 434,000 tonnes of apples into the UK, almost half of which, in 1996, came from non-EU sources.

While such data apparently makes little sense, it is important to bear in mind that the form of transport makes a difference to the environmental and social impact of freighting foodstuffs thousands of miles. The consultancy AEA Technology Environment, in preparing a report for the UK government on the ability of 'food miles' to indicate sustainable development, calculated the relative impact of food miles between transport modes. These calculations show the relative efficiency of sea transport which accounts for 65 per cent of the UK's food tonne kilometres, but only 12 per cent of its CO_2 emissions. Conversely, air transport accounts for only one per cent of food tonne kilometres, but 11 per cent of related CO_2 emissions. The environmental damage from aircraft emissions is even greater in practice, as greenhouse gas emission direct into the stratosphere is more damaging than at lower levels, as Chapter 8 on climate change shows (DEFRA, 2005).

Road haulage is the most damaging form of food transportation (33 per cent of food tonne kilometres but 57 per cent of CO_2 emissions are created by HGVs). This is even worse when consumers' own cars are taken into account for the journey between home and the food retailer. An assessment used by AEA of the direct impacts of food transport is provided below in Table 6.4.

Overwhelmingly, the largest component of foodstuffs that is air-freighted is vegetables (40%), followed by fruit (21%) and fish (7%). This kind of international trading rarely brings benefits of any substance to the producers. However, Suzanne Freidberg raises a number of related issues in her analysis of two post-colonial commodity networks, in which she argues the need to maintain appropriate world food trading networks, in order to maintain the fragile economies of poor countries, once colonies of European powers. She considers food imports, from Kenya to the UK and from Burkina Faso to France, in her attempt to answer key questions, such as where food should come from, whose food provisioning livelihoods we should care about, who should participate in food provisioning and through what kind of relationships, whether food should be sourced locally or globally, produced organically or scientifically and where the goodness in food lies (Freidberg, 2004). These are good questions for the reader to consider when trying to assess how food should be grown, transported and consumed.

Also recognising the damage that could be done to economies in the Global South if countries in the West stopped importing altogether, Bioregional Solutions, an organisation which develops practical solutions to environmental problems,[1] suggests ways in which exporters and importers need to be discriminating. While prevailing trade patterns need to be drastically overhauled, both to reduce CO_2 emissions and other noxious by-products of transportation, it is also wise to build independent economies in the less developed countries. Bioregional Solutions has developed a FEET (Foreign Exchange Earning per Transport Tonne of CO_2). Index which divides the amount of foreign exchange earned by the CO_2 released by transporting the product to the country of sale. A high FEET score represents high, and therefore desirable, foreign exchange earnings per unit of exported weight, whereas a low FEET score represents low earnings, and therefore undesirable, foreign exchange earnings per unit of weight. This index was used to calculate a range of products grown or manufactured in South Africa and imported into the UK, and found that grapes (8400 rand/tonne CO_2) and charcoal

Table 6.4   Impacts of food transport

Category	Type of Impact	Characteristics
Climate change	Environmental	CO_2 from transport fuel use (excludes refrigeration during transport)
Air quality	Environmental	NOx, SO_2, PM10, and VOCs
Noise	Environmental	Amenity costs (property prices, excludes health impacts)
Other		
Accidents	Social	
Animal welfare	Social	
Congestion	Economic	Quantitative assessment for road transport (partial for others)
Infrastructure	Economic	As above
Food prices	Wider	
Consumer choice	Wider	Includes nutrition and food culture
Rural communities	Other	Socio-economic issues and access to food
Developing countries	Other	Socio-economic and environmental issues

Source: DEFRA, 2005

(17,390) had a low FEET rating, but that wine (150,000) had an extremely high FEET rating. Bioregional Solutions' argument is that the UK should produce charcoal and fruit at home, and save the air miles for produce which is either difficult or impossible to grow locally, and that would earn its growers and manufacturers a better rate of return while causing less environmental damage (Desai and Riddlestone, 2002).

The DEFRA evaluation of food miles as an indicator of sustainable development in the UK (2005) has demonstrated the complexity of calculating the environmental damage of food miles, suggesting that 'a suite of indicators is needed to reflect the key adverse impacts of food transport'. For example, transport mode is important: while airfreight is highly polluting, 'large HGVs travelling long distances between suppliers and shops' can, through centralised distribution centres, enable very efficient loading of vehicles, compared to more local sourcing with smaller vehicles and lower load factors (DEFRA, 2005: np). Likewise, one case study found that it was 'more sustainable (at least in energy efficiency terms) to import tomatoes from Spain than to produce them in heated greenhouses in the UK outside the summer months'. Nevertheless, the writers of this report conclude that the 'increase in food transport has significant negative impacts on sustainability, including increased congestion, road infrastructure costs, pollution and greenhouse gas emissions', all of which they estimate generates over £9 billion in costs each year (DEFRA, 2005: 95).

While both Bioregional Solutions and Freidberg have examined inequalities between food exporters and importers at the global scale, the following section considers inequalities of food availability at both the global scale and within countries.

Inequalities of food availability

The Food and Agriculture Organization's (FAO) Special Programme for Food Security is committed to halving the number of the world's hungry by 2015; if it is able to do so, it will have achieved one of the Millennium Development Goals set by the United Nations in 2000. While this is a laudable aim, it has to be set against earlier attempts, such as the 1974 World Food Conference which pledged to eradicate global hunger within a decade (NAHO, 2004). Currently, according to the FAO, 825 million people are defined as being 'food insecure'. The FAO defines 'food security' as existing 'when all people, at all times, have access to sufficient, safe and nutritious food to meet their dietary needs and food preferences for a full and active life' (FAO, 2007). According to UNESCO, 25,000 people die every day of hunger. Mid-term progress reported by the United Nations for its Millennium Development Goals will be reviewed in the following section. The FAO has defined a minimum dietary energy requirement for countries in transition and developing countries, which ranges from 1720 kcals/day in the Lao People's Democratic Republic to 2030 kcals/day in the United Arab Emirates.

National calorie intake figures for Burundi (1685), Eritrea (1622), and Somalia (1566) fall below the lowest calculated minimum for individual consumption and a further 10 countries have an average per capita consumption of under 2000 kcals a day. It will come as no surprise that almost half of the children in these countries are recorded as being underweight (a measure that the United Nations uses as an indication of food shortage). Conversely, a number of countries consume more than 3500 kcals per head daily: Austria, Belgium, France, Greece, Ireland, Italy, Portugal, Turkey and the USA (Millstone and Lang, 2003).

These figures, of course, disguise a number of intra-national differences and there are clear differences in food consumption between men and women, by race and by income. In one of the world's richest countries, with one of the world's largest economies and one of the highest kcal pp/day intakes, the USA reports 11.1 per cent of its population (that is 12 million people) as being food insecure (Nord et al., 2002). The National Anti-Hunger Organizations alliance, in its 'Blueprint to End Hunger', published in 2004, challenged the USA to eliminate hunger nationally. It estimated that 35 million Americans were threatened with hunger, including 13 million children. Drawing on national data, a Massachusetts anti-hunger campaigning organisation, 'Project Bread' (2006), set out the links between food insecurity and poverty, in a state in which 7.1 per cent of households were estimated as food insecure and 2.7 per cent were reported as food.

When a family does not have enough money to buy food, one of two things can happen: they may end up with a virtually empty pantry and experience the pain of hunger, or they may try to fend off hunger with readily available, inexpensive, high calorie foods that have little or no nutritional value. (Alaimo et al., 2001; Townsend et al., 2001)

This situation is not confined to the USA, among Western countries. For a number of years, the Campaign for the Food Justice Strategies Bill has been working to pass legislation which would secure a legal duty on the UK Government to draw up and implement strategies to end food poverty[2]. The campaign cites evidence of up to 7 million people in the UK (that is, well over 10 per cent of the population) who cannot afford a healthy diet, and of up to 5000 people in each parliamentary constituency who either are, or are at risk of being, malnourished. One in four pregnant women are believed to be unable to afford a healthy diet, with the result that their babies are born with a low birth weight. Like Project Bread, reported above, the Campaign for the Food Justice Strategies Bill argues that childhood obesity affects 10 per cent of 6-year-olds and 17 per cent of 15-year-olds and that, for the same reasons (dependence on cheaper white bread, sugar and low fibre cereals, lack of access to fresh fruit and vegetables, living in what have become known as 'food deserts' where there are no supermarkets or greengrocers), children in low-income families are most at risk (CPAG, 2003). Obesity is further compounded by a lack of safe places to exercise, particularly in more disadvantaged neighbourhoods, a point that has recently been made by the Forestry Commission, 2005 and Project Bread, 2006. The last section of this chapter will review some initiatives that have been developed, largely by campaigning organisations, which provide low-income and otherwise disadvantaged households with fresh fruit and vegetables.

Given that the USA exports more food than it imports (US \$72.2 billion compared to US \$50.3 billion in 2000), and that this accounted for 13 per cent of global exports in 2000 (Millstone and Lang, 2003), that Americans are going hungry is clearly a food distribution issue, rather than a food shortage one. Drize and Sen's analysis of famine has already been referred to earlier in this chapter. Here they explain how, even in countries much poorer than the USA, famine situations usually affect only the poor in society, and are often caused by their inability to pay. Drize and Sen also demonstrates the distribution of food within the family, in which women are generally the last to eat, indicating that food accessibility is not only governed by income (Drize and Sen, 1989).

The role of international institutions

The previous sections have all demonstrated how the production and consumption of food creates and reinforces inequality at all scales and the impacts that this has both directly and indirectly on the environment. These last two sections review some attempts to mitigate these impacts. In this section, the role of international institutions concerned with controlling the worst impacts of food production and consumption on human health and the environment is considered. Given that food production and retailing is increasingly controlled by a small number of multinational companies, and given that more and more food is transported globally between the producers and consumers, as this chapter has already demonstrated, it is arguably necessary for governance over food to take place at the transnational and international scale. Lang and Heasman (2004) argue that although food regulations need to be enforced locally, the legislation itself is increasingly being made at the international level, through organisations such as the European Union (EU), the North American Free Trade Agreement (NAFTA) and the Asia-Pacific Economic Cooperation (APEC). Globally, the creation of the World Trade Organization (WTO) in 1994 has had a major impact on global food trading practices. The WTO incorporates a substantial number of subsidiary agreements in the General Agreement on Tariffs and Trade which, in general, reflect the lobbying ability of the USA and EU to protect their interests in national food production and biotechnology industries. The maintenance of selective farmer subsidies in the USA and EU effectively works against free trade for exporter countries, which prompted a coalition of these – the G-21 Group, led by Brazil, China and India – to demand that the USA and EU reduce their trade barriers, cut subsidies and open up markets to their cheaper commodities from elsewhere. The importance of this to poorer countries is reflected in the importance of agriculture to gross domestic product: in rich countries, it represents less than 2 per cent; in middle-income countries, 17 per cent; and in the poorest countries, 35 per cent (Lang and Heasman, 2004). Thus, the Cancún meeting in Mexico ended in stalemate, as rich and poorer countries were unable to find a compromise.

In the face of the power of the WTO, the World Bank and the International Monetary Fund, international agencies with a responsibility for health and the environment seem relatively powerless. As this chapter has already noted, the United Nations itself has declared the eradication of extreme poverty and hunger as one of its eight Millennium Goals, as well as ensuring environmental sustainability, both goals with a bearing on food. Specifically, the target regarding hunger is to halve the proportion of people who suffer from hunger between 1990 and 2015. By 2006 (more than halfway through the accounting period), chronic hunger (defined by the United Nations as insufficient food to meet daily needs) had declined proportionately but not enough to reduce the number of people going hungry, which increased to 824 million by 2003. This represented 17 per cent of developing regions' population, compared with 20 per cent in 1990–92. The situation in Sub Saharan Africa was particularly bad with 33 per cent of the population living with insufficient food in 1990–92, and 31 per cent in 2001–3 (United Nations, 2006).

One of the reasons why international legislation finds it difficult to meet targets such as these, is the lack of funding. Countries in the Global North have agreed (at successive conferences from the United Nations Conference on Environment and Development in 1992 to the G8 Summit of the World's most powerful economies at Gleneagles, Scotland in 2005) to increase their aid target to 0.7 per cent of gross national income. So far, only Denmark, Luxembourg, the Netherlands, Norway and Sweden have met this target. (United Nations, 2006).

From the brief analysis presented in this chapter so far, it seems unlikely that the current tendency to increased concentration of food production, and the inadequacies of national and international legislation, will deliver environmentally and socially sustainable food networks. Finally, this chapter considers the scope of small-scale projects to provide for local consumption needs.

Alternative food networks

Table 6.5   Categories of AFNs identified by Venn et al. (2006)

Category	Explanation	Examples
Consumers as producers	Food is grown by those who also consume it	Community gardens, community food cooperatives, allotment holders/gardeners
Producer–consumer partnerships	The risks and rewards of farming are shared between the food grower and consumer, either through shares or subscription	Community supported agriculture
Direct sell initiatives	Farmers sell direct to the consumer	Farmers' markets, farm gate sales, box schemes, farm shops
Specialist retailers	Shops selling high value-added quality or speciality food, including that which is locally produced	Online grocers, specialist shops

Broadly, these initiatives can be grouped under the banner of alternative food networks (AFNs). Venn et al. (2006) have reviewed an emerging body of work in this area and divide these networks into four categories, identified in Table 6.5.

This last section of the chapter will focus on two examples: farmers' markets and allotments and community gardens. Finally, the example of Cuba will be used to exemplify Lang and Heasman's third food paradigm: 'the ecological paradigm'. As an island, and as a result of a particular political ecology, Cuba has, since the early 1990s, transformed its food production system from one heavily dependent on artificial inputs and extensive trading links with the Soviet bloc, to one which is locally focused and organic. This is a useful way of illustrating both the potential of low-input food growing, and the importance of political ecology.

Glossary Box   Political ecology

Political ecology examines the relationship between society and nature, emphasising the role of political, social and economic factors in environmental change and vice versa. It developed out of a frustration with neo-Malthusian explanations of environmental problems in the Global South, usually involving concepts like 'over-population' for which non-democratic solutions were advocated, such as the infamous radios for vasectomies programmes in India in the 1960s and 1970s. Depending on who is doing the research, political ecology variously draws on class, gender, race and ethnicity examining how they intersect with local dynamics of place to construct and adapt environments (Bryant, 2001).

19 URBAN FARMING

AROUND 800 MILLION city dwellers worldwide – including some in industrialized countries – use their agricultural skills to feed themselves and their families. The world's cities are expanding at an ever-increasing rate. People are leaving agricultural regions no longer able to support them in order to find employment in urban areas. Some of the world's largest cities are now to be found in developing countries. However, most urban immigrants, do not find employment on arrival, but poverty and malnutrition.

As well as growing food for their own consumption, around 1200 million also earn a living growing food and rearing livestock to sell at local markets, while a further 150 million are employed as laborers on urban farms. The outskirts of most cities in Africa, Asia and Latin America support thousands of cattle, goats, pigs, chickens and rabbits, and both small and large livestock are also found in inner-city areas. When Hong Kong had an outbreak of avian flu in 1997, over one million chickens, housed in residential areas had to be destroyed – an unforeseen census.

More recently, as a result of the economic slow-down, in particular in East Asia, urban unemployment and poverty has risen. Increased urban food production is in many cases a response to these problems. In times of severe political or economic crisis, large cities are particularly vulnerable to food shortages.

During Indonesia's financial crisis in 1998, when food prices rose by 70 percent, the government encouraged people in Jakarta to grow food to prevent a breakdown of the fresh-food supply.

Supplying nutritionally adequate and safe food to city dwellers is a substantial challenge for governments and planners. Urban farming creates employment and income, increases food security and can improve the urban environment, but it faces still competition for land from developers. In order to ensure a healthy future for urban populations, policy-makers need a properly integrated policy that anticipates the changing needs of both rural and urban populations, and the links between them. Health controls are also important.

$500

million worth of fruit and vegetables is produced by urban farmers worldwide

Over 20 percent of Dar-es-Salaam, Tanzania is used for farming. Around 20 percent of workers do agricultural work, and 15,000 families in the city and its suburbs depend on income from agriculture. Urban farmers produce 90 percent of the leafy vegetables consumed in the city, and also raise livestock, including 6.5 million chickens. Some farms are beside rivers and have irrigation systems, but others rely on rainfall.

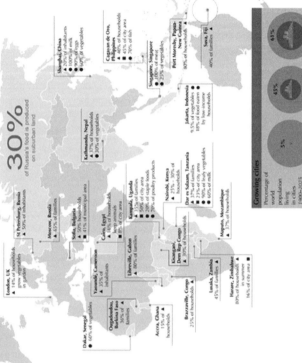

FOOD PRODUCTION IN URBAN AND SUBURBAN AREAS

▲ percentage of inhabitants or households involved in agriculture in urban or suburban area

■ percentage of city area used for agriculture

● percentage of agricultural needs of city met by urban and / or suburban production

30% of Russia's food is produced on suburban land

90,000 people in London rent small plots of land on which they grow vegetables and fruit.

London, UK
▲ 14% of households grow vegetables in garden

St Petersburg, Russia
▲ 50% of inhabitants

Moscow, Russia
▲ 65% of families

Sofia, Bulgaria
■ 41% of municipal area

Shanghai, China
▲ 20% of inhabitants
● 100% of milk
● 90% of eggs
● 90% of vegetables

Cagayan de Oro, Philippines
▲ 40% of households
■ 45% of city area
● 70% of fish

Kathmandu, Nepal
▲ 37% of households
● 30% of vegetables

Cairo, Egypt
▲ 16% of households keep large animals

Kampala, Uganda
▲ 35% of families
● 20% of staple foods
● 70% of poultry products

Singapore, Singapore
● 25% of vegetables

Jakarta, Indonesia
9.5% of vegetables
18% of food eaten by low-income households

Port Moresby, Papua-New Guinea
▲ 30% of households

Suva, Fiji
▲ 40% of families

Dakar, Senegal
● 60% of vegetables

Ouagadougou, Burkina Faso
▲ 36% of families

Accra, Ghana
▲ 15% of households

Yaoundé, Cameroon
▲ 35% of inhabitants

Libreville, Gabon
▲ 40% of families

Kisangani, Dem Rep Congo
● 80% of staple foods

Brazzaville, Congo
▲ 25% of households

Nairobi, Kenya
▲ 25% – 30% of households

Dar es Salaam, Tanzania
▲ 35% of families
■ 23% of city area
● 90% of leafy vegetables
● 60% of milk

Lusaka, Zambia
▲ 45% of households

Harare, Zimbabwe
80% of households
■ 16% of city area

Maputo, Mozambique
▲ 37% of households

La Paz, Bolivia
■ 30% of city area
● 30% of Cuba's requirements

Havana, Cuba
■ 41% of city area
● 30% of Cuba's vegetables

Mexico City, Mexico
▲ 1% employed in agriculture

Montevideo, Uruguay
● 6% of national pig production

Growing cities
Percentage of world population living in cities 1900–2025 projected

5% 1900

45% 1990s

61% 2025

Figure 6.1 Food production in urban and suburban areas from Millstone and Lang (© Myriad Editions/www.Myriad Editions.com)

Farmers' markets

Farmers' markets have long been common practice in the USA and these have inspired their more recent uptake in the UK. The first farmers' market to appear in the UK was in Bath in 1997, a joint initiative between Bath City Council and Bath Environmental Centre. In 1999, the National Association of Farmers' Markets was established by the Soil Association, Envolve, the Farm Retail Association and the National Farmers' Union. From 16 in 1999, there are now over five hundred farmers' markets in the UK, with 250 of these having official certification from the national association. Certification guarantees that the farmers' market is one in which farmers, growers or producers from a defined local area are present in person to sell their own produce, direct to the public. All products sold should have been grown, reared, caught, brewed, pickled, baked, smoked or processed by the stall-holder (National Association of Farmers' Markets, 2006). While this might be expected to reduce the food miles embodied in the produce traded, it may not always be the case. Wallgren's (2006) work in Sweden, a country with a short growing season, suggests that the wide catchment of Swedish farmers' markets, and the inefficient way in which produce is transported, means that there can actually be more food miles travelled. Nevertheless, she argues that with more efficient transportation (involving, for example, shared vans), this can be significantly improved.

Allotments and community gardens

Worldwide, 800 million city dwellers are estimated to grow their own food and 200 million are estimated to produce food for sale in local markets (Millstone and Lang, 2003). Figure 6.1 indicates the extent of food growing in a range of cities, which suggests a limited capacity for communities to secure food provision in the context of individual and national economic crisis. In addition to food security benefits, local food growing can make a positive environmental contribution. Jac Smit (2002) has argued that, in a Third World context, urban agriculture projects have increased civic stability, improved overall energy conservation, reduced traffic congestion and improved air quality, as well as giving enterprising, unskilled poor people a way to make money.

In the UK, there has been a long tradition of local food growing in the shape of the allotment, which Box 6.5 illustrates. The resurgence of interest in the British allotment has been evident alongside the development of community gardens, farmers' markets and other initiatives that Venn covers in Table 6.5.

Box 6.5   The British allotment

The British 'allotment' has its origins in issues of food poverty and social and economic inequality. Primarily a response to land enclosures of the 17th century, and related to rural to urban migration, the allotment provided small plots of urban land to households who would otherwise have lost the ability to provide their own food (Crouch and Ward, 1997).

(Continued)

Through the 20th century, allotments were in decline, apart from the two world wars, in which they were essential spaces for local food production at a time when it was not possible to import food. By the end of the 1960s, the majority of allotments were held by elderly, low-income men and the image of the allotment was quaint, old-fashioned and generally lacking in relevance to a late 20th-century world. However, with increasing concerns about food safety and more local commitment to sustainable development, allotments have seen something of a revival, with more women involved in growing food, more organically grown produce and a rise in demand in some areas. The National Society of Allotment and Leisure Gardeners (NSALG) reported 305,116 plots on 8025 sites, in 2001, representing an increase for the first time since 1970, when allotments started declining in the UK. This decline was both a result of demand for land for house building and a lack of demand for what was seen as an 'unfashionable' leisure pursuit. More recent concerns about chemicals used in food production, genetic modification of plants and more general issues such as the long haul of food creating more CO_2 emissions are encouraging younger people, and more women, to take up food growing. By 1993, the NSALG reported that 15% of their members were women (up from 3.2% in the late 1960s), and research undertaken by one of this chapter's authors in 2001 revealed that this was significantly higher in some areas such as the London boroughs of Hounslow (34%) and Richmond (41%). The same research found that women are more inclined to grow food organically, without recourse to pesticides and fertilisers.

(Buckingham, 2003, 2005)

Allotments offer a particularly powerful way for low-income communities to grow culturally relevant food at a fraction of the cost they would pay retail (and see earlier in the chapter for the paucity of fresh fruit and vegetables in low-income families' diets). While, as Figure 6.1 shows, the urban garden is a familiar feature of cities in developing countries, it has been reintroduced in North America, Europe and elsewhere in the West. This has largely been accomplished by environmental, health and anti-poverty activists and NGOs, as both an anti-poverty device and a means for improving the environment. Food security needs to be recognised as more than simply an absence of hunger, as the earlier discussion on health pointed out. Gottleib and Fisher (1996) argue that it should include the availability of '. . . a culturally acceptable, nutritionally adequate diet through local, non-emergency sources . . .' which may include 'urban greening initiatives such as community gardens [and] edible landscape plantings . . .' (p. 196). One example (see Figure 6.2) of how this can be provided is Tucson Urban Gardens in Arizona, USA, which have been created to enable low-income, mostly Hispanic, families to grow food, and where the surplus is sold at cost through the adjoining food bank supermarket. Plots are gardened by local people, but also by youths who have been excluded from school. Associated programmes involve the collection of product surpluses notified to the project by farmers and other gardeners and, through this, youths are also developing food preparation skills, as they learn to make lemon curd and pecan pies, which are then sold in the on-site supermarket (Buckingham, 2003).

Figure 6.2   Tucson Urban Gardens (Reproduced with permission from Buckingham et al., 2003)

Other inspiring examples of community gardening include rooftop gardening in St Petersburg which has been experimented with to try to overcome shortages of basic foodstuffs (to the extent that, as Figure 6.1 shows, 50 per cent of St Petersburg residents grow some of their food). Using roofs built to support heavy burdens of snow in the winter, it is estimated that two thousand tonnes of vegetables may be grown in a year. Vegetables grown on rooftops have also demonstrated health benefits, as they were shown to have less heavy metal contamination than those grown at ground level in the city (Gavrilov, 1996).

As Figure 6.2 has already illustrated, community gardens are an increasing feature of schools which are committed to environmental education, and also contribute to one of the aims of Agenda 21, to include children and young people in environmental decision making. Teachers in West London, interviewed by a third-year student undertaking research into school community gardens, suggested that a wide range of subjects can be taught, ranging from literacy (reading seed packet instructions and writing gardening diaries), numeracy (measuring plots), design and technology (building the plots), geography (identifying how different plants come from different areas), science (ecosystems, weather and soil conditions) and environmental issues, such as recycling. The Black Environmental Network (BEN) in the UK has also worked with a number of schools in multi-ethnic areas to develop 'cultural garden projects', in which students can learn about each other's cultural heritage from the food different communities eat.

Growing food in cities provides ample opportunities for waste recycling, both as an outlet for organic waste that can be used for compost, and for materials such as plastic bottles for cloches to protect seedlings, or old carpets which can be used to cover soil to prevent weeds growing. Third World urban agricultural gardening has been reported as using more recycled material than common rural food production methods, as well as fewer inputs (such as water and chemicals). This and the fact that it produces a greater variety and higher quality of food, and between three and fifteen times the yield per hectare, suggests that there may be benefits to expanding urban agriculture elsewhere (Smit, 2002). Indeed, Dongtan, a new 'eco-city' in China designed to operate ecologically in a closed loop, plans to recycle all of its organic waste as compost for food growing that will take place in the surrounding farmland (Girardet, 2006).

Cuba

A rather different example of local and organic food growing is in Cuba. Following 'perestroika' in the Soviet Union in the late 1980s, Cuba lost its major trading partner and communist ally. Instead of following this political lead, Cuba declared a 'special period' in which socialism was reaffirmed, although it recognised an increasing need to engage with foreign capital. Geraldine Lievesley, in her book on the Cuban revolution shows how Cuba became increasingly self-sufficient in food in the 1990s. 'Before the break up of the soviet bloc, 57 per cent of the calories in the Cuban population's diet had come from imported food . . . . Between 1989 and 1993, calorific intake was reduced by 33 per cent' (Lievesley, 2004: 163). This was in part compensated by food rationing which ensured equitable distribution, and special provision was made for children. Emerging out of this situation, however, what began as a survival strategy to produce more locally grown food without the artificial inputs previously supplied by the soviet bloc, has become something of a template for other countries. Not to be confused with the 'Green Revolution' (p.127), 'the greening of the Revolution' has seen a shift from large-scale industrial farming to a combination of traditional knowledge and modern ecological science. An organic farming programme was launched in 1991, and the Cuban army was despatched to work on farms which were being transformed from state enterprises to basic production units and cooperatives. For the first time since the revolution in 1958, individuals were allowed to grow their own fruit and vegetables in towns and cities and these *huertos* (patio gardens using only natural pesticides), together with the *organoponicos* (state-owned gardens on land previously urban wasteland), were producing 60 to 70 per cent of Cuba's vegetables by 2004 (Lievesley, 2004). Through its involvement in international fair trade, Cuban cooperatives produce oranges and other fruit for juice drinks sold just above the current world price, which yields a premium which is then invested in development and welfare work. According to Lievesley, 'the results of this farming revolution have been spectacular, ending Havana's dependence upon food brought in from the countryside and ensuring that Cuban dietary habits have changed for the better as people eat far more fruit and vegetables' (2004: 165). The demise of artificial chemicals has also resulted in increased biodiversity on the island, which has in turn contributed to an increase in ecotourism, itself an earner of foreign exchange.

Summary

As the introduction to this book makes clear, social relations are embedded in environmental issues. Food is an excellent example of this, and what this chapter has sought to demonstrate is that the scientific, economic, political and social mechanisms through which food is produced, distributed and consumed have very specific, and uneven, environmental and social effects which it is impossible to disentangle.

Despite decades of programmes ostensibly designed to increase the food supply, the world has more hungry people now than ever before. The irony of this is that at the same time, the wealthy West has a higher than ever proportion of its population suffering from problems of consuming too much, and the wrong kind of, food.

Using the case study of food also illustrates the interconnections between environmental problems. The chapter makes clear that issues concerning food penetrate other environmental dilemmas, such as waste and climate change, and that it is impossible to try to solve any one in isolation. Doing so will not only have negative environmental effects elsewhere, but social effects also.

Notes

1   Examples of these practical solutions include burning local biomass waste for local energy, converting suburban wasteland to lavender farming by offenders, and the development of the internationally acclaimed BedZED – Beddington Zero Energy Development home/work community in South London.
2   The Campaign for the Food Justice Strategies Bill is a coalition of twenty national organisations campaigning for food justice. It also tabled an 'Early Day Motion (No. 70)' at the beginning of the 2005 Parliamentary session which needs the support of 20 Members of Parliament to be admitted for debate in Parliament. This has been done each year since 2002, with the aim of securing enough votes. See www.foodjustice.org.uk

References

Adams, B. (1998) *Timescapes of Modernity*. London: Routledge.
Alaimo, K., Olson, C.M. and Frongillo, E.A. (2001) 'Low family income and food insufficiency in relation to overweight in children: is there a paradox?', Archives of Pediatrics and Adolescent Medicine.
Bryant, R. (2001) 'Political ecology: a critical agenda for change?', in N. Castree and B. Braun (eds), *Social Nature: Theory, Practice and Politics*. Oxford: Blackwell.
Buckingham, S. (2003) 'Allotments and community gardens: a DIY approach to environmental sustainability' in S. Buckingham and K. Theobald (eds), *Local Environmental Sustainability*. Cambridge: Woodhead Publishing.
Buckingham, S. (2005) 'Women (re)construct the plot: the regen(d)eration of urban food growing', *Area* 37(2): 171–9.
Carranza, A., Domingo A. and Estrades, A. (2006) 'Pelagic longlines: a threat to sea turtles in the equatorial Eastern Atlantic', *Biological Conservation*, 131: 52–7.
Carson, R. (1962) *Silent Spring*. Boston: Houghton Mifflin.

Child Poverty Action Group (CPAG) (2003) 'Campaign for the Food Justice Strategies Bill' Campaigns Newsletter. http://www.cpag.org.uk/campaigns/CPAGCampaignNews June2003.pdf

Crouch, D. and Ward, C. (1997) *The Allotment, its Landscape and Culture*. Nottingham: Five Leaves.

DEFRA (2005) *The Validity of Food Miles as an Indicator of Sustainable Development*. London: DEFRA.

Desai, P. and Riddlestone, S. (2002) *Bioregional Solutions for Living on one Planet*. Dorcester: Schumacher Briefings/Green Books.

Drèze, J. and Sen, A. (1989) *Hunger and Public Action*. Oxford: Clarendon Press.

Ethical Consumer Magazine (2006) http://www.ethicalconsumer.org/

FAO (Food and Agriculture Organization of the United Nations) (2007) http://www. fao.org/righttofood/ (accessed 9 August 2007).

FSA (Food Standards Agency) (2003) www.food.gov.uk/healthiereating

Forestry Commission (2005) 'Economic benefits of accessible green spaces for physical and mental health: scoping study', *Final Report for the Forestry Commission*. Oxford: CJC Consulting.

Freidberg, S. (2004) *French Beans and Food Scares: Culture and Commerce in an Anxious Age*. Oxford: Oxford University Press.

Gabizon, S. (1998) 'A dying sea and a dying people', in *Women, Health and Environment*. Utrecht, The Netherlands: Women in Europe for a Common Future.

Gavrilov (1996) plant@sovam.com www.cityfarmer.org/russiastp.html

Girardet, H. (2006) 'Which way China?', *Resurgence*, 236: 18–19.

Gottleib, R. and Fisher, A. (1996) '"First feed the face": environmental justice and community food security', *Antipode*, 28(2): 193–205.

Greenpeace (2006) http://oceans.greenpeace.org/en/footer-links/search-results?q=plastics+ vortex (accessed 7 December 2006).

Guardian Unlimited (2006) 'Are Supermarkets Sweeping Up?' November 6, 2006.

ICES (2006) International Council for the Exploration of the Sea. http://www.ices.dk (accessed 11 October 2006).

Klein, N. (2001) *No Logo*. London: Flamingo.

Lang, T. and Heasman, M. (2004) *Food Wars: the Global Battle for Mouths, Minds and Markets*. London: Earthscan.

Lapierre, D. and Moro, J. (2002) *Five Past Midnight in Bhopal*. New York: Warner Books.

Levin, P.S., Holmes, E.E., Piner, K.R. and Harvey, C.J. (2005) 'Shifts in a Pacific Ocean fish assemblage: the potential influence of exploitation', *Conservation Biology*, 20: 1181–90.

Lievesley, G. (2004) *The Cuban Revolution: Past Present and Future Perspectives*. Basingstoke: Palgrave, Macmillan.

Lucas, C. (2001) 'Stopping the great food swap: relocalising Europe's food supply'. Brussels: The Greens/European Free Alliance in the European Parliament.

Millstone, E. and Lang, T. (2003) *The Atlas of Food: Who Eats What, Where and Why*. London: Earthscan.

National Anti Hunger Organizations (2004) 'A Blueprint to End Hunger', National Anti Hunger Organizations; http://www.centeronhunger.org/pdf/Blueprint%20final.pdf

National Association of Farmers' Markets (2006) http://www.farmersmarkets.net/ (accessed 24 November 2006).

National Statistics (2006) North Sea Fish Stocks. http://www.statistics.gov.uk/ (accessed 17 October 2006).

Nord, M., Andrews, M. and Carlson, S (2002) 'Household food security in the United States'. Food and Rural Economics Division, Economic Research Service, US Department of Agriculture, Food Assistance and Nutrition Research Report No. 35.

Popkin, B.M. (1999) 'Urbanization, lifestyle changes and the nutrition transition', *World Development*, 27(11): 1905–16.

Project Bread (2004) 'The link between hunger and obesity', Boston, MA. http://www.projectbread.org/site/DocServer/TheLinkBetweenHungerAndObesity_2004.pdf?docID=104

Project Bread (2006) 'Status report on hunger in Massachusetts' Boston, MA. http://www.projectbread.org/site/DocServer/StatusReportOnHungerInMA_2006.pdf?docID=621

Shiva, V. (2002) *The Sir Albert Howard Memorial Lecture*. Soil Association. www.soilassociation.org/web/sa/saweb.nsf (then search for Shiva, accessed 16 April 2004).

Siegle, L. (2006) 'One family, one month, 50 kg of packaging: why?', *The Observer Magazine*, 29 January.

Smit, J. (2002) 'Urban agriculture: a powerful engine for sustainable cities', in D. Taylor-Ide and C.E. Taylor, *Just and Lasting Change: When Communities own their Futures*. Baltimore, MD: Johns Hopkins Press.

Townsend, M., Peerson, J., Love, B., Achterberg, C. and Murphy, S. (2001) 'Food insecurity is positively related to overweight in women', *Journal of Nutrition*. 131: 1738–45.

UNDP (2003) United Nations Human Development Report at http://hdr.undp.org/statistics/data/indicators.cfm?x=21&y=1&z=1 (accessed 1 February 2006).

United Nations (2006) Prevalence of undernourishment in total population (percentage): Food and Agriculture Organization of the United Nations. http://www.fao.org/faostat/foodsecurity/Files/PrevalenceUndernourishment_en.xls

Venn, L., Kneafsey M., Holloway L., Cox R., Dowler E. and Tuomainen, H. (2006) 'Researching European "alternative food networks: some methodological considerations"', *Area*, 38(3): 248–58.

Wallgren, C. (2006) 'Local or global food markets: a comparison of energy use for transport', *Local Environment*, 11(2): 233–51.

Wrap (2005) 'Major Retailers join WRAP in pledging to tackle packaging and foowaste'. http://www.wrap.org.uk/retail/news_events/news/major_retailers.html

7 Waste

Susan Buckingham and Iris Turner

Learning outcomes

Knowledge and understanding:
- of how waste is socially defined, and how it is becoming increasingly challenged as a necessary by-product of our lives and consumption patterns
- as well as learning about current mitigation measures, you will also learn how the way in which waste is disposed of is socially inequitable at a range of scales from the local to the international

Critical awareness and evaluation:
- you are invited to reflect on the incompatibility between a high consumption lifestyle, promoted as an aspiration in advanced capitalist societies, and this socio-environmental inequity
- finally, you might like to reflect on your own role in contributing to the problem, as well as how you make changes in your own life (at home, in university and elsewhere) to help solve the problem

Introduction

Think of your average day: you get up and wipe off the residue of last night's make-up with a disposable wipe, or shave with a disposable razor. Too late for breakfast, you buy a take-out latte grande in a disposable cup, or maybe a plastic bottle of water or coke and a nutri-bar, and bring these to your lecture. The cup, bottle and wrapper are tossed into the rubbish bin at the end of the lecture. For lunch, you self-serve yourself a salad or stir fry at the university refectory into a plastic container, grab another bottled drink, paper napkin, individually wrapped salt and pepper, and sit down to fish the newspaper you bought (made from 75% recycled paper) out of its plastic bag. With eyes bigger than your stomach, you scrape your left-over food into the general rubbish before heading off for the afternoon's sessions. The assignment you're handing in later has been printed off a few times – the (unrecycled) toner was clearly on the way out and half your print was unclear, and then you found a careless error on page 4 – single-sided on high-grade virgin pulp. Your lecturer (responding to popular demand) has printed out copies of the PowerPoint presentation she's about to give, and extracts from a book she wants you to read and discuss next week.

All the above constitute but a small proportion of the waste generated by a student on a modest income in a rich Western country in the early 21st century. Add to this the packaging of the food, toiletries, cosmetics and other goods we buy and the products which have come to the end of their – to us – useful, or fashionable, life, and it is no surprise that material outputs to the environment from economic activity in five study countries investigated by the World Resources Institute range from 11 metric tonnes per person per year in Japan to 25 metric tonnes per person per year in the USA. When non-domestic flows are included, total material outputs to the environment range from 21 metric tonnes per person in Japan to 86 metric tonnes per person in the USA. The same report identified that one half to three-quarters of annual resource inputs to industrial economies are returned to the environment as wastes within a year (Matthews et al., 2000: xi).

Western society has been defined as a 'disposable' or 'throwaway' society – from the once-used nappies and wipes to the appliances which it is no longer financially viable to repair; goods and materials are thrown away without a thought about implications for the environment, or the communities who are most exposed to the means of disposal: the landfill sites and incinerators. This chapter will look first at what constitutes waste, how waste is defined in different places and times, the different kinds of waste produced in Western societies, and the problems these present, through a review of the hierarchy of waste treatment from waste reduction to waste disposal. It will consider the communities which are most exposed to the pollution and health problems arising from waste disposal, and their relationship with NIMBYs (Not in my Back Yarders) who use their wealth, educational privilege and relative political clout to avoid the negative consequences of their high-consumption lifestyle. Whether looking at poor communities within Western societies, or countries in the Global South pressured to accept the waste of high consumption, these are powerful examples of environmental injustice, a concept already explained in Chapters 2 and 3. Finally, the chapter will explore arguments that the only viable way to address both the social and environmental problems associated with a high consumption, high waste production and disposability society is to engage with the possibility of achieving 'zero waste'.

Defining waste

Box 7.1   Defining waste

'It is characteristic of all life that it takes in suitable raw materials, e.g. food and air, and converts them into products of value to itself or its species, e.g. heat, energy, body material, progeny. In doing so, it inevitably produces waste material, e.g. carbon dioxide, faecal matter, which it

(Continued)

---

must get rid of or perish. In the special case of modern human life the 'intake' is of much wider character, and includes also fuel, clothes and the general appurtenances of civilisation. Our waste is correspondingly large and varied, and includes air pollutants, water pollutants, and the contents of dustbins. So far, a higher standard of living has always been accompanied by a larger volume of more complicated waste, and there is no sign that this will not be true in the future. This waste must be disposed of. Industry is no exception to this rule. It uses raw materials, processes them to yield useful products, and is left with waste which may exceed 50% of the raw materials used. It is quite inevitable, and as industry increases in extent and diversity so will the waste increase in amount and complexity. It is a liability and must be disposed of.'

Source: Ministry of Housing and Local Government/Scottish Development Department, 1970

Table 7.1 Types of waste produced from different sources in the UK

	Type of Waste	Nature of Waste
Controlled Waste	Municipal	Any materials from sources such as households, street litter, and some shops and offices.
	Industrial	Waste generated by industrial plants, factories and storage facilities.
	Clinical waste	All human or animal bodily tissues, fluids or excretions; pharmaceutical products; dressings; sharp implements; any other waste arising from medical, dental, veterinary or pharmaceutical practices.
	Commercial	Waste from commercial establishments like shops, offices and catering which are responsible for their own waste disposal.
	Hazardous	Any waste which might cause harm to human health, animals, plants and the environment. The most common sources are industrial chemical processing, metal processing, oil refining, solvents, waste oils and asbestos.
	Radioactive	See Box 7.2.
Uncontrolled Waste	Agricultural	Waste from a farm or garden nursery, including chemicals, organic matter and packaging.
	Mining and quarrying	Any waste from mining and quarrying activities, but also including waste from building and demolition works, such as bricks, concrete timber and hard core.

Source: DETR, 2000a; Environment Agency, 2007a

The remit of the 1970 commission from which the quotation in Box 7.1 was taken was to address the disposal of solid toxic waste and the resulting report represents an important point in the development of the UK's waste management strategy. The UK has, however, performed worse than many European states in the amount of waste it produces and in the minimising of the environmental impacts of waste disposal. The quotation also emphasises the fact that wastes are produced by all human activities. As the Earth's resources are limited, it is vital that every effort is made to re-use or find an alternative use for waste by direct or indirect recycling of these materials, rather than disposing of them in landfill sites or by incineration.

The EU's Waste Framework Directive 2006/12 defines waste as 'any substance or object in the categories set out in Annex I which the holder discards or is required to discard.' Waste can be defined in a number of ways, including where it is generated (see Table 7.1), its environmental impact, its form (whether it is solid, liquid, gaseous, slurry, or powder), its properties (whether it is toxic, reactive, acidic, alkaline, inert, volatile, carcinogenic), its legal definitions (for example, special, controlled, household, industrial), and ways in which it is disposed of. This chapter focuses on waste defined by source, and by method of disposal.

The Environment Agency regulates controlled wastes, as indicated in Table 7.1, including household, industrial and commercial wastes. Other wastes, arising from agricultural activities, mining and quarry operations, are non-controlled wastes and are not yet regulated by the Environment Agency. Radioactive waste is a special case and is subject to separate controls and its nature and disposal is dealt with in some depth in Box 7.2 below.

Box 7.2   Radioactive wastes

Radioactivity is the spontaneous emission of radiation, either directly from the decay of unstable atomic nuclei or as a consequence of a nuclear reaction. During decay, new substances are formed which may also be radioactive. Since they decay indefinitely, with the number of atoms steadily decreasing, the rate of decay of radioactive substances is shown by their 'half-life', which is the time it takes for half of the atoms of the substance to decay. Half-lives can vary from a few milliseconds to several million years for different substances, however, the half-life of any one substance is always the same. The shorter the half-life, the faster the radioactive substance decays and the more unstable it is. The longer the half-life, the slower the decay process and the more stable the substance. The final product of the decay process produces stable non-radioactive substances. People are exposed to radiation from a range of sources including rocks and soils, from which 19% of the Earth's radioactivity is received; cosmic radioactivity from outer space (14%); X-rays (11.5%) and air travel (0.5%). The greatest concern about radioactivity, however, is related to nuclear power generation.

Radioactive waste comprises any substance arising from any process involving radioactive materials. The differences in treatment of different types of radioactive wastes depend on the properties of radioactive materials. Radioactive wastes are classified, according to their activity, heat generation potential and physical characteristics, as *low-level*, *intermediate-level*, or *transuranic* waste and *high-level* wastes. (Nuclear Information and Resources Service, 1992). Low-level wastes constitute the bulk of radioactive waste.

(Continued)

Exposure to radioactivity can cause a number of harmful health effects, which are related to the period of exposure, dosage and radiation type. Short high-level exposure causes radiation sickness and temporary sterility, while long-term, low-level exposure may produce no physical effects for decades. Radiation sickness is also known as acute radiation syndrome (ARS), immediate symptoms of which include nausea, vomiting and diarrhoea. Longer term, exposed populations may develop bone marrow depletion leading to weight loss, flu-like symptoms and bleeding. Lower doses of radiation accumulate over time, with risks of cancer, cataracts and decreased fertility. Radiation injuries can be divided into two types: somatic, which damage the irradiated person; and genetic, where the offspring of the irradiated person is affected by radiation damaging the germ cells in the reproductive organs (Southern Cross University, 2005). For these reasons, it is essential that high-level radioactive wastes are isolated from the environment and human populations over periods of thousands of years.

The problems of disposal have not been solved, particularly for high-level waste. However, low-level wastes do not generally need any processing prior to disposal in shallow land burial. For the most part, intermediate wastes need to be incorporated into inert materials such as concrete, bitumen or resins, before being stored at designated sites awaiting decisions on suitable longer-term disposal methods. High-level wastes are stored initially in containers placed in water-cooled ponds, in order to dissipate the heat generated by the radioactive isotopes before their subsequent disposal. All intermediate and high-level waste must be stored at depth.

The storage of certain radioactive wastes is still very problematic since the timescale for protecting humans may be millions of years. In the meantime, the wastes must be kept isolated from their immediate environment at all times. As CoRWM (the Committee for Radioactive Waste Management) has concluded, the only viable long-term disposal method for high-level radioactive wastes is to seal them into a suitable inert medium and then place them in a deep underground facility, pending further research on effective disposal (CoRWM, 2006). In the UK, this facility is located at Sellafield in Cumbria, while the USA is preparing to encapsulate all its high-level civil spent fuel assemblies in an underground storage facility in the Yucca Mountains, Nevada. Other, technically more advanced methods are also being employed, for example, in Australia using a type of synthetic rock for disposal (Hore-Lacy, 2003; World Nuclear Association, 2007).

Industrialisation has introduced a vast range of new products and materials into the waste stream, from the point of extraction of the raw materials used to make the product, through manufacture and distribution, to disposal of the product once it has reached the end of its useful life. Analysing the environmental impact of this stream of activities as 'life cycle' or 'cradle-to-grave' analysis is outlined in the UK Government's Waste Strategy, and will be considered later in the chapter (see Box 7.3). Increasingly, appliances have shorter lives and households often have more than one of each appliance. For example, in the UK, the relatively new technology of DVD players and recorders escalated from sales of 18,000 a year in 1984 to 6.8 million in 1994. The average number of TVs currently in a UK home is now 2.5, just one of the estimated average 25 electrical and electronic appliances owned by householders in the UK, representing an increase of 60% between 2000 and 2005 (Holdway, 2005).

The bulk and composition of domestic waste has changed over time, related to social and economic shifts in styles of living. Of particular importance has been the rise of supermarkets and increased competition for sales, which has led to an emphasis on the marketing of goods accompanied by a sharp rise in packaging, as Chapter 6 illustrates with respect to food. Despite a significant increase in recycled packaging waste (56 per cent in the UK by 2006, according to Defra, 2007c), it is still a significant part of the domestic waste stream and needs to be reduced. Despite the search for less environmentally damaging packaging materials than plastics (such as biodegradable and photodegradable materials) there are some practical problems to overcome before they can effectively replace conventional plastics (Waste Watch, 2004). Currently there are a number of concerns about the use of degradable plastics, including possible methane emissions, and the notion that there may be an increase in waste and litter if the public think that discarded plastics incorporating organic compounds will just disappear.

Packaging legislation attempts to reduce the amount of plastic material which ends up in the waste stream. The UK Government's national packaging recycling and recovery targets specify that 24.5 per cent of plastic waste is recycled by 2008 (Letsrecycle.com, 2007).

Of course, one household's waste is potentially a resource for another. There are an increasing number of schemes by which prosperous households can dispose of items, such as furniture and clothes, to organisations such as the Salvation Army, for recycling to homeless households moving into settled accommodation, or to charity shops. In addition, households can earn income from selling on goods through car boot fairs and eBay, while organisations arrange to collect the computers that companies replace with the latest model, for recycling to less well resourced charities, or to businesses in the Global South. These examples of reuse and circulation of goods will be developed later in the chapter. A number of opinion formers, such as the Royal Society of Arts and Greenpeace, are advocating a 'zero-waste society' in which all waste is treated as a resource (Murray, 2002) and this chapter will conclude by reviewing some of these proposals.

Waste disposal

In the past, depositing domestic wastes in a hole in the ground was accepted, since it was cheap and space was readily available in old mines and quarries. Possible environmental effects such as noise, dust, odours, traffic and the pollution of surface and ground waters were considered unavoidable consequences. The sites available for landfill in the UK, are however, diminishing rapidly and the Environment Agency estimates that within 5 to 10 years, there will be no suitable spaces left (Environment Agency, 2007b). At the same time, the environmental impacts of waste disposal are becoming increasingly unacceptable. One way of attempting to identify these environmental impacts and to determine the effectiveness of waste disposal is 'life-cycle' or 'cradle-to-grave' analysis of products, which Box 7.3 explains, along with some of their complexities. However, it is increasingly being argued that life-cycle analysis is not enough, and that a 'cradle-to-cradle' assessment needs to be achieved.

Box 7.3   Analysing environmental impact

Life-Cycle Analysis (LCA) is a way of examining the total environmental impact of, for example, a manufactured product through every stage of its life from: '... the mining of the raw materials all the way through its manufacturing, sale, use, re-use and, finally, disposal. LCAs enable a manufacturer to quantify how much energy and raw materials are used, and how much solid, liquid and gaseous waste is generated, at each stage of the product's life. Such a study would normally ignore second-generation impacts, such as the energy required to fire the bricks used to build the kilns used to manufacture the raw material . . .

'. . . Recycling introduces a further real difficulty into the calculations. In the case of materials like steel and aluminium that can technically be recycled an infinite number of times (with some melt losses), there is no longer a 'grave'. And in the case of paper, which can be reprocessed four or five times before fibres are too short to have viable strength, should calculations assume that it *will* be recycled four times, or not? What return rates, for example, should be assumed for factory-refillable containers? . . . and materials sent for recycling. The transport distance in each specific case is a major influence on the environmental impacts associated with the process. An LCA, which concludes that recycling of low-value renewable materials in one city is environmentally preferable, may not hold good for a different, more remote city where reprocessing facilities incur large transport impacts . . . In most situations it is impossible to prove conclusively using LCAs that any one product or any one process is better in general terms than any other, since many parameters cannot be simplified to the degree necessary to reach such a conclusion . . .

'. . . When first conceived, it was predicted that LCA would enable definitive judgements to be made. That misplaced belief has now been discredited. In combination with the trend towards more open disclosure of environmental information by companies, and the desire by consumers to be guided towards the least harmful purchases, the LCA is a vital tool.'

(DETR, 2000b; World Resource Foundation, undated)

Cradle-to-cradle analysis

The concept 'cradle-to-cradle' was developed by Michael Braungart and William McDonough to describe the circulation of resources through one of two cycles, either biological (where everything grown from the land returns to the land, free from toxic, persistent and bio-accumulative contamination), or technical, whereby materials are constantly recycled (IPPR and Green Alliance, 2006: 12). Braungart's own website advertises 'cradle-to-cradle certification' as providing 'a company with a means to tangibly, credibly measure achievement in environmentally intelligent design and helps customers purchase and specify products that are pursuing a broader definition of quality'.

'This means using environmentally safe and healthy materials; designing for material reutilization, such as recycling or composting; the use of renewable energy and energy efficiency; efficient use of water, and maximum water quality associated with production; and instituting strategies for social responsibility. If a candidate product achieves the necessary criteria, it is certified as a Silver, Gold or Platinum product or as a Technical/ Biological Nutrient (available for homogeneous materials or less complex products), and can be branded as cradle-to-cradle.'

Source: Braungart (undated)

The complexity of the environmental impact is also identified by Holdway (2005), in relation to the increased number of electronic products referred to earlier. Holdway has identified over 5 million TVs being disposed of annually in the UK, 'many to landfill, despite the fact that they contain a potentially hazardous cocktail of materials. Cathode ray tube displays contain between 4 lb and 8 lb of lead and most of the solder used in circuit boards is also lead . . . a neurotoxin which is harmful to kidneys and nervous and reproduction systems, and which inhibits the mental development of young children and foetuses. CRTs also contain phosphor and barium, short-term exposure to [which] can cause brain swelling, muscle weakness and damage to the heart, liver and spleen' (Holdway, 2005: 25). The landfilling of TVs and other electronic goods is now illegal in the UK under the Hazardous Waste Regulations introduced in 2004, in the EU, and in a number of US states.

At the United Nations Conference on the Environment and Development (UNCED) in 1992, Agenda 21 signatories committed to a waste-free society, and the UK Government pledged to recycle 25 per cent of rubbish by 2002, although this target has not yet been achieved across the UK. This commitment was reinforced at the World Summit on Sustainable Development in 2002 and the European Union has legislated a programme of waste reduction, setting stringent targets for the recycling of different types of materials. A sustainable waste management strategy is characterised by a waste hierarchy shown in Table 7.2, and it is expected that this will, in order of priority, be worked through before resorting to incineration or landfill. Sustainable waste management is defined as 'using material resources efficiently, to cut down on the amount of waste produced and where waste is produced, dealing with it in a way that actively contributes to the economic, social and environmental goals of sustainable development' (United Nations, 2001). Such a strategy would be expected to involve effective protection of the environment, a sensible use of natural resources, the conversion of waste to a resource, and social progress which meets the needs of everyone, economic growth and employment.

Table 7.2   Waste hierarchy

Strategy	Example
Reduction	Reduced consumption and unpackaged goods.
Re-use	The life of an object is extended by using it in a different way or by circulation through, for example, charity shops, car boot sales, eBay/Freecycle, LETS.
Repair	For example, clothes, electrical goods, furniture.
Recovery (recycling, composting, energy recovery)	Community schemes and local authority collections. (Some governments include incineration, where this includes heat recovery.)
Incineration	Without heat recovery.
Landfill	This can be authorised and unauthorised (e.g. fly-tipping).

Source: Prime Minister's Strategy Unit, 2002; SITA, 2004; Waste Watch, no source date

The rest of this section will continue to use the UK as a case study, placing it within the European context. Because different countries classify waste in different ways, it is difficult to compare data; however, Figure 7.1 illustrates the differences in performance between the fifteen EU countries before the 12 accession states, mainly from Central and Eastern Europe, joined between 2005 and 2007. Although there is no standardised method of categorising data across the member states, it is clear that Greece, Portugal and the UK are heavily dependent on landfill for the disposal of their wastes and perform poorly in recycling. However, this needs to be put into context, as while Greece and Portugal generated 438 and 446 kilograms of waste per person, respectively, in 2005, the UK generated 584 kpp (Eurostat, 2006).

The main waste disposal strategies will be considered in the order of their environmental impact, starting with landfill.

Landfill

As Figure 7.1 shows, of the material sent for waste disposal in England and Wales, nearly 75% of it goes to landfill, a fast diminishing facility, particularly around built-up areas. Domestic waste can produce polluting liquids and explosive gases which, when buried in the ground, can contaminate surface and ground waters. Although it is practical to site landfill facilities close to urban areas, where much of the domestic waste is produced, this is not popular with many residents, who do not want such sites near their home. As a result, waste materials have to be transported further from their source, generating pollution in the process. There are other constraints on locating landfill sites, for example, they should not be too close to either surface or underground water, the geological and hydrological conditions should be stable, and sites sensitive because of their natural or cultural heritage must be avoided. Disposal of domestic wastes to landfill causes a number of environmental concerns, which are shown in Table 7.3.

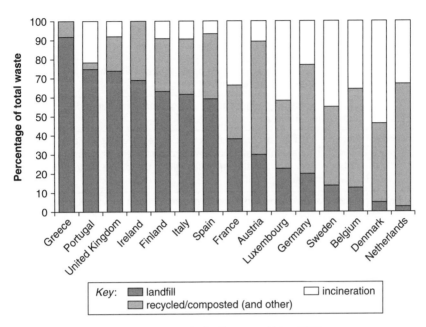

Figure 7.1 Municipal waste management in the European Union 2003

Source: Eurostat, 2006

The EU Landfill Directive 1999/31/EC which came into force in the UK in 2001, and is implemented through the Pollution and Prevention Act 1999 and the Landfill (England and Wales) Regulations 2002, aims to reduce the adverse environmental impacts of landfill sites, and restricts or prevents the landfilling of certain types of waste. In addition, it imposes targets for reductions in the amount of biodegradable domestic waste which can be landfilled, so that by 2010, the maximum amount of waste to landfill should be 35% and biodegradable municipal waste going to landfill must be progressively reduced. By 2010, this should be reduced to 75% of the total produced in 1995; by 2013, to 50%; and by 2020 to 35% (The Council of the European Union, 1999). In addition, the Directive requires specific controls on the disposal of hazardous waste and bans tyres completely from landfill by 2006 (Letsrecycle.com, 2007a). To encourage these targets to be met, the UK Government imposed a landfill tax escalator which is progressively increasing landfill tax to £35 per tonne by 2012 (HM Treasury, 2003). Arguably, such taxes, and any potential tax on domestic waste disposal, may encourage illegal waste dumping, known as fly-tipping, which is becoming a significant problem in any vacant spaces that are not easily monitored. Materials fly-tipped can range from green waste to domestic items, to abandoned cars and construction waste, much of which may be hazardous or toxic, not to mention unsightly. The 2003 Clean Neighbourhood Act introduced tough new powers for the British courts to take effective action against fly-tippers, although local authorities find it difficult to identify fly-tippers and often lack the resources to take action.

Table 7.3   Environmental concerns of landfilling

Media	Effect
Water	Leachate containing a number of possible chemical compounds with a number of toxic metals such as lead and mercury.
Soil	Leachate containing a number of chemical compounds including nitrates, phosphates and lead.
Atmosphere	Methane gas from decomposing organic material; odour and dust emissions; wind-borne materials; traffic pollution through transporting waste such as carbon dioxide (CO_2), nitrogen oxides (NO_x), particulate matter (PM), volatile organic compounds (VOCs).
Flora/Fauna	Attracts vermin, birds and insects which might infect humans with diseases such as Weil's disease.
Other	Noise from transport; danger of fire; site security; reduces aesthetic quality.

Source: DETR, 2000c

Incineration

For thousands of years, the value of burning wastes has been recognised, both to reduce the quantity of waste generated by households, trades, and agricultural practices, and to provide fuel for heating or cooking. Recognition of the potential environmental problems generated by burning wastes also has a long history. In the UK, the existence of city controls on the burning of rubbish in open dumps can be traced back to the 13th century (Project Integra, 2000). The Industrial Revolution, and accompanying rural-to-urban population migration of the 18th and 19th centuries, transformed the nature and volume of wastes arising and the potential health problems of improper disposal practices. Mass burning of wastes in enclosed and controlled conditions became an important waste management option (Chartered Institution of Water and Environmental Management, 2006).

Incineration plants heat the wastes to high temperatures in order to decompose the waste materials, ideally to safe end products. However, depending on the nature of the wastes, the end products are likely to include a number of gases, metals and particulates, some of which are known to be hazardous.

Despite the versatility of incineration as a waste treatment method, opposition developed to such an extent that in many countries, including Britain, proposals for new incineration plants faced long delays and often refusal of planning permission. From the 1970s, a number of incinerator plants were decommissioned amid public concern about their safety. Of particular concern have been the emissions of organic compounds, which arise from the incomplete combustion of organic wastes present in the original waste, and the synthesis of 'new' compounds. Such materials contain a range of toxic chemicals, including the family of dioxins which are known to be extremely toxic to both animals and humans, even in small amounts (Health Canada, 2006), although these materials can be destroyed when heated to a very high temperature, likely to be in excess of 1200 °C (Chartered Institution of Water and Environmental Management, 2006; Cheung et al., 2007; Knox, 2005).

By the end of the 20th century, incineration technology had improved substantially, so that emissions of gaseous materials and small dust-size particles could be contained within the plant itself, and were no longer considered to be a serious health threat. By 2003, the Environment Agency was advocating incineration as the best practicable environmental option (BPEO) and the UK Parliamentary Office of Science and Technology supported a steep increase in the number of incinerators (Chartered Institution of Water and Environmental Management, 2006; The Parliamentary Office of Science and Technology, 2000). Nonetheless, despite the improvements, there remain health and environmental concerns related to atmospheric and waste-water discharges. Incinerator plants are becoming more popular across the European Union and in other parts of the world (Market Research.com, 2006), although getting planning permission can pose difficulties if they are to be sited near large populations.

The specific advantages of incineration are that it reduces the original volume and weight of waste, especially of bulky solids with a high combustible content; reductions achieved can be up to 90 per cent of the volume and 75 per cent of weight of materials going to final landfill. Another advantage is that the destruction of some wastes and detoxification of others render them more suitable for final disposal (Wasteonline, no source date). The organic component of biodegradable waste can also be destroyed which would otherwise generate methane gas if sent to landfill. The energy recovery potential of incineration has made it increasingly attractive to municipalities charged with reducing their use of fossil fuels and production of carbon dioxide. These factors have led to the range of waste incinerated expanding in many industrialised countries.

The use of domestic wastes as a fuel to produce energy is a positive application of its removal. Energy is recovered from incineration in the form of heat, which is used to convert high-pressure steam into electricity. Probably the most well known scheme currently in the UK is in Woking, Surrey, where the local authority set up and owns an energy company, generating heat and power in a network of sixty local generators (combining co-generators and tri-generators with photo-voltaic arrays and a hydrogen fuel cell station). The resulting heat and power is communicated to all council buildings in the centre of town, as well as social housing and local businesses via a 'private wire'. The scheme is groundbreaking because of the carbon dioxide emission reductions achieved (77 per cent in 15 years), the local ownership of the generators and the social benefits it delivers. For example, households in fuel poverty have made significant savings in their fuel bills (Greenpeace, 2005).

These applications of previously 'waste' heat and energy from incineration plants has led to their being designated as recycling facilities, and, indeed, the energy generated is rather disingenuously currently classified as renewable energy by the UK Government. However, there is a range of problems associated with incineration's increased use as part of an integrated waste management programme. For incinerators to work effectively, they need to operate at full capacity over a sustained period – there is, then, little incentive for municipalities to reduce the waste stream if this will lead to interruptions in the supply of material to the incineration plant. So while the health impacts from new plants are substantially less than previous generations of incinerators, this 'high technology' response to the problem of excess waste faced by high consumption societies is still seen as problematic by environmental campaigning groups (Friends of the Earth, 2007). Also, as suggested earlier, the power and influence of wealthy communities enables them to successfully campaign against incinerators 'in their backyard', resulting in an environmental justice issue where poorer communities (in the UK and USA) and non-white communities in the USA are more likely to be the recipients of incinerators and their negative side effects, as the section on waste and equitability will later demonstrate.

Recovery

Waste recovery is defined as the use of a material not necessarily in its original form. It involves the collection and separation of materials from waste and subsequent processing to produce marketable products, such as paper and board, aluminium cans, compost or soil conditioner, fuel (solid or gaseous) and energy (heat or electricity), as described above. Some recycled products bear little resemblance to their previous life: for example, fleeces made from recycled plastic.

Recovery falls into three classes:

1   Re-use is the highest form of recycling in that it requires the least energy and it is the simplest process to prepare an item for re-use, such as refilling compressed gas cylinders.
2   Direct recycling is the intermediate form of recycling. In this class, more energy is used and the process of recycling is more complicated than re-use. For example, a glass bottle no longer suitable for re-use in its original form will be broken down, melted and then made into another bottle.
3   Indirect recycling is the lowest form of recycling in that an item will be recycled in a different form. The process is more complex and more energy-intensive. For example, a glass bottle no longer suitable for re-use or direct recycling will be ground down to produce a material, such as for road surfacing (Porteous, 1996).

Recycling offers many advantages, including the conservation of natural resources, reducing the demand for incineration and landfill space, reducing the demand for both energy and virgin raw materials, producing goods more cheaply, and also giving people who recycle an individual sense of responsibility concerning the waste they produce. This can, however, divert people from minimising their waste by making them feel that they are already making their contribution to waste minimisation. Defra data reveals that, while recycling rates have increased steadily since 1983, overall per capita waste has risen over this time (Defra, 2007d).

The responsibility for recovery of waste lies with individual waste management authorities (in the UK, a complex arrangement exists between waste collection and waste disposal authorities), and this is regulated by environmental agencies for England, Wales, Scotland and Northern Ireland. There is considerable variance in recycling performance between waste management authorities, even within London, where recycling rates vary from 8.8% in Tower Hamlets to 37.7% in Bexley for the year ending in December 2006 (Defra, 2007a). In the period 2005/2006, the average recycling rate, including composting, for all the London Boroughs was 20.7% (Defra, 2007a; Sills, 2007), while for the whole of England it was 26.7% (Defra, 2007b). Clearly, there is a need to increase recycling rates substantially in order to reach targets both for the UK and the EU. For all the waste materials recycled, the locations for treatment of these materials are often some considerable distance from the recycling point, for example, much of London's waste paper is sent to Aylesford in Kent or Ellesmere Port in Cheshire (Ollie Recycles, no source date).

In order to get the general public to participate in the recycling of their wastes, education is required from primary-school level onward. Many primary schools collect and separate their waste materials, some of which are then used in the classrooms and school grounds. In respect of educating adults, a number of strategies have been proposed to increase both recycling and reduce conventional waste, from rewarding appropriate behaviour with financial incentives, and offering prizes, to an intensive education programme to promote participation in waste reduction and recycling schemes (AEA Technology Environment, 2006; Prime Minister's Strategy Unit, 2002). Furthermore, as is commonplace in Europe, a variety of taxes and levies are currently under consideration as incentives for recycling materials in the UK (Defra, 2007c). The Waste and Resources Action Programme (WRAP) has been set up by the UK Government to promote sustainable waste management.

Table 7.4 which follows presents a range of materials that are typically recovered, either in kerbside collections or at 'bring' sites, and their key recovery characteristics.

Up to 60 per cent of household waste can be biodegraded by composting (Scottish Executive, 2006), hence it makes sense to reduce the amount of material going to landfill, much of which would otherwise produce methane, by getting domestic households to recycle their organic material on-site. Essentially, composting is the process by which biodegradable material, such as kitchen and garden waste, is decomposed. This process generates considerable heat (the normal range is 25°– 40 °C) which breaks down what was once waste into a stable and nutrient-rich soil conditioner.

The main purposes and advantages of composting are the stabilisation of waste, the destruction of pathenogenic bacteria and viruses (through the intense heat generated) and nutrient reclamation. In their complex organic forms, the nutrients nitrogen, phosphorus and potassium are difficult for plants to absorb. However, after composting, these nutrients are converted into inorganic forms such as nitrate and phosphate, which are more easily absorbed by plants. Compost may be used as a soil conditioner to improve soil structure, enhance biological activity and as a growing medium for the horticultural industry. Above all, compost reduces the need to supply organic materials to the soil from elsewhere, such as the addition of peat, which causes a significant loss of natural organic material that would take many thousands of years to replace.

Table 7.4   Recylable goods and their properties

Material	Qualities
Paper	Different types of paper need different amounts of processing, and are suitable for different types of end product, from new paper at one extreme, to packaging and boards at the other. Also, repeated recycling means that the range of applications eventually becomes limited.
Cardboard	Cardboard is made entirely from recycled material. Apart from the obvious uses, it has a wide range of applications, from the construction industry to furniture.
Oil	All waste mineral oils contain traces of additives and other contaminants. The disposal of oil waste is regulated by the European Union Waste Oil Directive.The oils can be cleaned up in a variety of ways, from laundering for re-use, to combustion.
Batteries	Automotive batteries must be dismantled at nominated recovery sites where hazardous materials such as acid, lead and other environmentally damaging metals can be extracted for re-use. New materials can also be generated from the other parts of batteries. Currently, Britain sends all consumer batteries to France for recycling.
Electrical and electronic goods	The European WEEE (Waste Electrical and Electronic Equipment) Directive, in force since 2003, requires the collection, treating, recycling and recovery of associated waste. Further directives specify how hazardous waste in electrical and electronic equipment, such as precious and environmentally hazardous metals, must be dealt with. Also, since 2000, ozone-depleting substances such as chlorofluorocarbons (CFCs) and hydrochlorofluorocarbons (HCFCs) must be removed from refrigeration and air-conditioning equipment and safely disposed of.
Glass	Glass can be recycled indefinitely, but it can become contaminated with foreign objects or by mixing with different coloured glasses. Bottles are the first product from the recycling of glass. Subsequent recycling is likely to produce cullet (broken glass) which is used for aggregate in the construction industry and 'glassphalt' for road building.
Aluminium	The metal has a number of physical properties that enable it to be recycled numerous times without deterioration. Aluminium is ideal for food packaging and has a large number of applications in industrial and domestic settings.
Steel	Steel has magnetic properties, so it is easily separated from other metals in recycling. It can be recycled many times without deterioration of its physical characteristics.
Plastics	Only plastic which melts on heating (thermoplastics) can be recycled by remoulding. Bottles normally made from three polymer types, namely: (a) polyethylene terephthalate (PET) for fizzy drink bottles; (b) high-density polyethylene (HDPE) for milk and detergent bottles and (c) polyvinylchloride (PVC) for large squash bottles, can be recycled several times. Recycled plastics can also be used as an alternative to wood in many applications, as it is cost-effective, durable (including in harsh and wet conditions) and does not leach any harmful chemicals into the soil or water. In Fowey Harbour, Cornwall, more than 1000 metres of recycled plastic has been utilised around the port as pontoons, walkways and fencing as a replacement for the traditional wood.
Fluorescent tubes	Require a specialist recovery procedure as they contain mercury, a neurotoxin and carcinogen, which seriously affects human health, and

Table 7.4 *(Continued)*

Material	Qualities
	is environmentally contaminating. They also contain polychlorinated biphenyls (PCBs) which must be incinerated.
Textiles and shoes	Textiles and shoes are made from a variety of synthetic and natural materials, and if they are in good condition, they can be re-used. Otherwise, they are sorted by fibre type and can be recycled by a variety of processes. There is a wide variety of end products, including yarn for new cloth, paper and floor tiles.
Toner cartridges	These items include rubber, foam, plastics and paper. The cartridges can be re-assembled and filled several times after dismantling.
Paint	Paint contains a range of chemicals, exposure to which can lead to a variety of health problems. Hence, many paints are considered to be hazardous and special arrangements need to be made for their collection and subsequent disposal as they are very toxic. Small amounts of household paints can be forwarded on to community organisations.
Scrap metals	These are derived from either new scrap (from metal processing activities, such as off-cuts, stampings, turnings, grinding and swarf from industry), or old scrap (from end-of-life or obsolete products, dismantling plants or processing of consumer goods).
Tyres	Abandoned tyres may spontaneously combust and the smoke is highly polluting. Motor vehicle tyres (whole or shredded) are banned from landfill sites in the UK to avoid this. It may be possible to re-use or retread some whole tyres. Alternatively, the tyres may be shredded in order to make new objects, such as roof tiles and bike pedals. Recently, tyres have been used to make tyre-derived fuel which is more efficient than some more traditional fuels.
Wood	From a recycling perspective, wood has the disadvantage of biodegrading unless painted or varnished. Wood for recycling can be classified according to its origins: (a) green waste; (b) untreated timber; (c) structural wood waste normally coated with a chemical as a preservative; (d) wood processing activities such as sawdust and (e) manufactured goods. Hence, the application of recycled wood is diverse, ranging from structural elements of a building, such as roof rafters, to kitchen utensils such as plates and bowls.

Source: DETR, 2000a

Because composting converts a waste product into a useful substance, it is classified as a waste recovery operation under the Waste Framework Directive 75/442/EEC (Wasteonline, no source date). At the present time in the UK, organic wastes are mostly collected from civic recycling sites, and from public parks and gardens, with only a small percentage collected at the kerbside. However, the UK Government has set targets for composting, which include recycling 33 per cent of domestic organic material by the year 2015 (Letsrecycle.com, 2005).

As the information above suggests, much of the UK legislation on waste recovery is driven by European legislation. The UK performs quite weakly compared to many European countries, particularly those in Northern Europe, as Figure 7.1 has already indicated. However, it would be a mistake to assume that high recycling rates denote low waste creation, as has already been referred to. One UK waste management authority, which was a case study for a European project on the gender mainstreaming of waste (see later), recorded relatively high recycling rates (at 22 per cent, well above the national average), and yet also had one of England's highest increases in conventional waste generation (Buckingham et al., 2005). Arguably, as Watson and Bulkeley (2005) have noted, 'encouraging residents to separate their waste for a kerbside scheme is a widening of horizons from waste disposal', but this does nothing to change their practices of consumption, which generates the waste – recycled or incinerated. Indeed, some campaigners (such as the Women's Environmental Network) argue that recycling can encourage residents to feel that they are doing their bit for the environment, so that they do not question or, therefore, modify their high-consumption lifestyle. Partly as a result of these campaigns, there is increasing academic and community attention being paid to re-use and repair of materials, which not only has an environmental benefit of reducing materials and energy use, but also has social benefits. Alternative circulation networks, through which goods cost less and enable low-income households to acquire clothes, furniture and other goods at reduced prices, are explored further below.

While circulating goods through alternative networks has a long history (apart from exchanges between family and friends, think of jumble sales, bring-and-buy stalls and so on), the mechanisms through which this exchange now takes place have proliferated in ways which reflect the changing nature of our society, in which networks are as likely to be facilitated by information technology as through the local community.

Charity shops make a positive contribution to the recycling of books, toys, clothes, tools, kitchenware, bric-à-brac and small items of furniture donated by individuals. Car boot sales in the UK have developed as a means by which individuals and households can raise money by selling goods they no longer want, and the Internet has facilitated a range of recycling opportunities, both for profit (eBay) and without charge (Freecycle). The increase in all of these trading mechanisms also makes these goods available at low cost for people with low incomes.

'Give and take' days are organised by local authorities, and environmental and community groups to which residents can bring along a variety of items they no longer need, to exchange for those they do. LETS (Local Economic Trading Schemes) can also be considered here, as they provide a mechanism for community members to circulate goods and services using an alternative currency (Aldridge et al., 2003; Seyfang, 2003). Gregson and Beale have also researched the circulation of clothing, noting its particular prevalence in maternity wear and children's clothes. When all these issues are considered, it is clear that waste, in terms of what is discarded, or in Gregson and Beale's (2004) term 'ridded', is far from being a straightforward environmental matter, but one which also relates to identity. This suggests that the ways in which society deals with waste need to be tackled more broadly than through technical solutions, hitherto the mainstay of waste management.

Another re-use strategy which has been promoted by the Real Nappy Network, and is now taken up by many local authorities in the UK, is to encourage new parents to use washable nappies rather than the disposable nappies heavily marketed by their manufacturers in neo-natal units. Disposable nappies constitute 3 per cent of waste going to landfill in the UK and, while the Environment Agency argued in a controversial report published in 2006 that washable nappies carry their own environmental burden of water, heating and detergent use, environmental campaigning groups maintain that the cradle-to-grave impact of disposal nappies is greater than laundering reusables. Some local authorities are subsidising reusable nappies as a poverty reduction strategy to relieve low-income households of the expense of continual purchase of disposable nappies. This chapter will return to re-use when considering zero-waste strategies in its conclusion.

Watson and Bulkeley (2005) have argued that the structure of waste legislation in Europe is such that there is little incentive for waste authorities to reduce waste at source. The emphasis on recycling as an alternative to incineration and landfill merely provides a different way for individuals and households, and waste management authorities, to dispose of the waste produced. It fails to challenge conventional product development or consumption practices. They argue that a 'move towards focusing on the underlying causes of the distributional problems of waste' (p.423) could create greater environmental and social equity, as less waste would need to be disposed of. In the last two sections of this chapter, the concepts of environmental equity and a zero-waste society will be explored in more depth.

Waste and equitability

The amount of waste produced by households, and by societies, bears a direct relation to the amount those households or societies consume, and this, in turn, is linked to their material wealth. Such re-use and recycling is governed by thrift occasioned by necessity (see Chapter 10 on Mexico City): a poor household is more likely to buy clothes from a charity shop, throw out clothes when they are threadbare rather than when they are out of fashion, replace electrical goods as they break and generally buy fewer non-essentials. In short, a poor household is likely to have a smaller 'ecological footprint' than a richer one. There is, however, another dimension to inequity with regard to waste, which is linked to its disposal. The environmental justice movement in the USA, as Chapter 3 has already discussed, was galvanised by the frustration of poor black communities who were more likely to receive a toxic waste dump in their backyard than white communities. Checker (2005) presents evidence to argue that race in the USA is a predictor of whether or not a census tract hosts a toxic waste facility, and 60 per cent of African-Americans and Hispanics, and 50 per cent of Asian/Pacific Islanders and native Americans, live in communities with at least one uncontrolled toxic waste site. The Center for Policy Alternatives in the USA reported that there was an 8 per cent increase in the degree to which race predicted the siting of hazardous waste facilities (that is, more than poverty, land values or home ownership) between 1987 and 1993. Checker (2005) reports that delays in listing hazardous sites in minority areas on the National Priority List were 20 per cent longer than those in predominantly white areas, and that penalties charged to polluters in white areas were 500 per cent higher than in minority areas. This, coupled with the circumstantial evidence of disproportionate ill-health suffered by minority populations in the USA, whereby African-Americans, for example, are three times more likely to die from asthma than white Americans, illustrates how waste has become a potent factor in mobilising environmental justice campaigns.

Walker et al. (2003) concluded from a study of integrated pollution control (IPC) sites in England that waste facilities had the strongest spatial correlations with areas of deprivation, while in Scotland, Dunnion and Scandrett (2003) have reported on a community outside Glasgow, Greengairs, which takes waste from the city as landfill. They have argued that it is the industrial dereliction and high unemployment, resulting from previous generations of industrial exploitation such as mining, that has enabled the area to be planned for waste disposal. In Greengairs, however, it was not only waste from Scotland that was dumped, but also soil contaminated with polychlorinated biphenyls (PCBs) exported from Hertfordshire (a relatively prosperous county in southern England). While the legislation governing the landfill company's site in Hertfordshire prohibited it from accepting PCBs, failure in Scottish environmental legislation enabled the company to use its Scottish site. As well as affecting a poor, working-class community with health and environmental hazards, Dunnion and Scandrett argue that the peripheral economic and geographical location of Scotland within the UK makes it more vulnerable to environmental exploitation.

There is, then, plenty of evidence linking poor and otherwise disadvantaged communities with noxious activities, such as waste disposal plants, although a note of caution is sounded by Judith Petts (2005) who, in a review of a number of epidemiological studies, suggests that it is difficult to find conclusive evidence linking the siting of waste facilities and ill-health. This is for a number of reasons, including the difficulties of separating out location, poverty and/or race from other issues such as individuals' lifestyle and exposure to other environmental problems. Petts also argues that the benefits of proximity to waste facilities are under-explored and cites the potential advantages of jobs, or access to 'energy-to-waste' projects which provide low-cost heat and power (Petts, 2005).

With regard to municipal waste, the majority of waste management decisions are made by a relatively privileged group of technocrats: trained engineers whose default solutions to the waste problems identified are engineering-based. Research into the gender mainstreaming of waste for the European Union explored the ways in which women were involved in waste management: from the household level (in terms of who decided what was to be recycled, for example), to policymaking. Waste management, unsurprisingly, was found to be a highly masculinised activity, grounded in high-technology, engineering-based solutions. This was in contrast to the structuring of waste management decisions in the home, which tend to be dominated by women (for example, whether, how and what to recycle; whether to buy heavily packaged or unpackaged goods). It was, therefore, not surprising that waste management authorities in which women had made inroads into the profession, tended to have higher success rates in increasing recycling. The evidence collected suggested that this was both because they had entered waste management from alternative starting positions, which were more likely to include education (Ireland) and re-use (West Sussex), and that they often understood the basis from which women residents made their waste management decisions (Buckingham et al., 2005).

Inequalities in the exposure to pollution linked to waste disposal also occur at the broader scale between the Global North and South, as prosperous countries and companies in the North export waste to countries in the South for processing and/or disposal. While the export of hazardous waste has been controlled since 1992 by the Basel Convention on the Control of Transboundary Movements of Hazardous Wastes and their Disposal, there continues to be a significant amount of both legal and illegal shipment of hazardous waste. The Convention estimates that 8.5 million tonnes of hazardous waste is shipped internationally, each year (Basel Convention, 2005). The Convention requires that all exported waste defined as hazardous, including chemicals such as arsenic, lead and mercury, bio-medical waste, bioaccumulative waste (i.e. persistent organic pollutants) and PCBs, has 'prior written notification' from the exporting state, and is treated in ways to minimise health and environmental risk. The Basel Action Network has condemned the Convention as 'an instrument that served more to legitimize hazardous waste trade rather than to prohibit what many felt was a criminal activity' (Basel Action Network, 2007), and campaigned for a complete ban on the international movement of hazardous material. In 1998, six years after the Convention came into force, 'The Basel Ban' was agreed, which bans the export of hazardous waste from OECD countries (the twenty-nine most wealthy coun-tries in the world) to non-OECD countries, although not without considerable opposition from the USA, Canada, Australia and South Korea (Basel Action Network, 2007). In 2003, end-of-life ships were designated as hazardous waste, which requires them to be decontaminated before exported for final dismantling.

Greenpeace, which monitors the effectiveness of the ban, has reported many contraventions of the Convention, however, and is sceptical of how transparent the shipping industry is. Its website details a number of past and ongoing trans-gressions, which is well illustrated by the case of the 'Probo Koala'. This ship, chartered by a shipping company, and flying under the Panamanian flag (a 'flag of convenience' which does not indicate the owner of the ship – a bit like the way in which cold-callers suppress their telephone numbers so that we cannot trace the call) was carrying mercaptens and hydrogen sulphide – both highly noxious-smelling toxic chemicals. The Probo Koala made a first attempt to unload its cargo in Amsterdam, but the expense of having the waste disposed of in this OECD country provoked the ship to try to find alternative, cheaper waste disposal options, and it ended up in the Côte d'Ivoire (Ivory Coast). It was only after six people in Abidjan, the capital, died, and almost 9000 sought medical attention, that this came to the attention of the international community, which sent a mis-sion (comprising the World Health Organization, the United Nations Disaster Assessment and Coordination and a team from France) to investigate further (Greenpeace International, 2006).

This example, as others on the Greenpeace International database, illustrates a dilemma for poor countries in the Global South, which offer cut-rate waste disposal options, achieved by low or absent health and safety, and environment, regulations, in order to earn foreign exchange. The catalogue of breaches of the Basel Convention are testament to a global case of environmental injustice. The Convention itself advocates a 'three-step strategy', which includes the minimisa-tion of the generation of wastes, as well as treating waste as close to its point of generation as possible, and reducing international movements of international waste (Basel Convention, 2005). In a zero-waste society, to which this chapter finally turns, this particular environmental injustice would not be able to persist.

Zero-waste society

Advocates of a 'zero-waste society' (ZWS) argue that society needs to stop viewing the waste stream as a linear process in which waste is the end of the process, and begin seeing it as a circular system in which materials are increasingly smartly used, thereby reducing waste, and transformed. Not only will this make environmental good sense, but economic good sense also. This is, of course, not a new approach to resource use and it characterises pre-industrial societies. Nineteenth-century American transcendentalist writer Ralph Waldo Emerson wrote: 'Nothing in nature is exhausted in its first use. When a thing has served its end to the uttermost, it is wholly new for an ulterior service' (Hawken et al., 1999: 17).

This was implicit in *Factor Four*, written by Ernst von Weizsacker et al. (1995), which was shorthand for doubling wealth and halving resource use, and which sought to demonstrate the financial gains of resource efficiency. Building on this, a subsequent publication by Hawken, Lovins and Lovins in 1999, *Natural Capitalism*, advocated harnessing these 'eco-efficiencies' to 'a fundamental rethinking of the structure and the reward system of commerce' (Hawken et al., 1999: x) to avoid both material and human waste.

In his book, *Zero Waste*, Robin Murray suggests that there are three key drivers for zero waste: pollution from waste disposal, climate change and resource depletion (2002: 5). As the previous section has indicated, waste management has hitherto been a specialist industry with its own technocratic experts. To move towards a ZWS, waste needs to be put into the wider context of the manufacture and use of products. This involves more intelligent product design, including the phasing out of materials which cannot safely be disposed of. Murray gives examples of product chains, cycles and, ultimately, spirals, in which the aim is to achieve 'upcycling' that not only conserves materials from one product to another, but that adds value in the recycling process. He uses the example of rice husks which originally 'posed a waste disposal problem in Asia because they were incombustible.' Alternative uses have been found for them, including as 'a substitute for polystyrene as a packaging material for electronic goods and then, after that use, as a fire-resistant building material' (Murray, 2002: 27). This example comes from the work of Michael Braungart's model of a service economy, in which products are designed so that, if they 'cannot degrade back into natural nutrient cycles', they can 'be deconstructed and completely reincorporated into technical nutrient cycles of industry' (Hawken et al., 1999: 17–18). Finding alternatives to hazardous materials, reducing consumption, smarter manufacturing and effective 'upcycling', as exemplified above, can all, Murray argues, result in zero atmospheric damage, zero toxic discharge and zero material waste.

The concept of ZWM, as used by Murray, emerged from Total Quality Management (TQM) in Japan, one of whose defining concepts was 'zero defects'. The practice of this has led to extraordinarily low rates of defects, for example, one defect per million claimed by Toshiba. As Japanese industry extended this to aspiring towards zero waste, Honda (Canada) reduced its waste by 98 per cent in a decade (Murray, 2002: 19).

New Zealand has committed the country to achieving zero waste by 2020, although no data is available on what gains have been made so far (IPPR and Green Alliance, 2006: 28). An increasing number of municipalities have committed to a goal of ZWM for example, Canberra, Australia by 2010 which had achieved 73 per cent of waste recycled by 2006; (IPPR and Green Alliance, 2006: 28); Kamikatsu, Japan (by 2020, and had achieved between 75 and 80 per cent by 2006, IPPR and Green Alliance, 2006: 28); and San Francisco (also by 2020, with 75 per cent diverted from landfill by 2010: by 2006, the city had achieved a 67 per cent recycling rate). Chew Magna in Somerset, England has developed a 'Go Zero' campaign, promoting strategies such as renewable energy, a shuttle commuter bus, local food growing and buying local, the circulation of unwanted goods through a 'swap it' noticeboard and 'clear your house out' days (http://www.gozero.org.uk/). This clearly links to some of the concepts introduced earlier in the book, such as ecological footprinting (Chapter 1) and bioregions (Chapter 2).

Murray's book was commissioned by Greenpeace and, as one of the world's most significant environmental campaigning organisations, it is not surprising that it is advocating zero waste. What, perhaps, is more striking, is the uptake of the concept by more mainstream organisations, such as the Institute for Public Policy Research and the RSA (Royal Society for Arts, Manufactures and Commerce). The IPPR report, which the Green Alliance was commissioned to produce, proposed a 'cradle-to-cradle' evaluation of a product's impacts, which presupposed that each product was a point on a continuous cycle of material recycling. Like Hawken et al. (1999), the report calls for a re-emphasis of a product on the service it provides, 'rather than on the ownership of the product itself' (IPPR and Green Alliance, 2006: 6). The report goes on to argue that the current waste agenda, focused on recycling targets, reducing packaging and diverting waste from landfill is stagnating and 'urgently needs more life' (ibid.: 6), to which end it makes a series of recommendations for minimising waste through incentives and charging, encouraging smarter, less wasteful product design and promoting cradle-to-cradle evaluation.

'Moving towards a zero-waste society' is one of the five manifesto goals of the RSA, which it is doing with a series of 'art and ecology' projects which raise awareness through creative communication. Its objectives for 2006–7 included the launch of a personal carbon trading scheme, running educational projects, and holding international workshops and conferences, primarily through creative engagement (see www.RSAartsandecology.org.uk).

Summary

In this chapter, we have moved from conceptualising waste as an end product which has, in some way, to be dealt with, to considering it as part of a cycle of material production. While it is now recognised, in the West, that landfill is no longer a tenable waste option, recycling must also be seen as only one option, less desirable than reducing consumption. Considerations of environmental damage and unequal social impacts demand that our consumption, recycling and disposal practices be carefully examined and revised.

References

Aldridge, T., Patterson, A., and Tooke, J. (2003) 'Trading places: geography and the role of LETS in local sustainable development, in S. Buckingham and K. Theobald (eds), *Local Environmental Sustainability*. Cambridge: Woodhead Publishing.

Basel Action Network (2007) What is the Basel Ban, http://www.ban.org/about_basel_ban/what_is_basel_ban.html (accessed 1 August 2007).

Basel Convention (2005) *Minimising Hazardous Wastes: A Simplified Guide to the Basel Convention*. Chatelaines, Switzerland: Basel Convention/UNEP.

Braungart, M. (undated) 'Cradle to cradle certification'. www.braungart.com/c2c_certification.htm

Buckingham, S., Reeves, D. and Batchelor, A. (2005) 'Wasting women: the environmental justice of including women in municipal waste management', *Local Environment*, 10:4.

Chartered Institution of Water and Environmental Management (2006) 'Waste incineration', www.ciwem.org./policy/factsheets/wastes.asp (accessed 24 March 2007).

Checker, M. (2005) *Polluted Promises: Environmental Racism and the Search for Justice in a Southern Town*. New York: New York University Press.

Cheung, W.H., Lee, V.K. and McKay, G. (2007) 'Minimising dioxin emissions from integrated MSW thermal treatment', *Environ. Sci. Technol.*, 41(6): 2001–7.

CoRWM (Committee on Radioactive Waste Management) (2006) 'Deciding the future of the UK's radioactive waste', www.Corwm.org.uk (accessed 1 April 2007).

Defra (2007a) 'London's recycling rates', www.defra.gov.uk/environment/statistics/wastekf/wrfk07.htm (accessed 17 July 2007).

Defra (2007b) 'e-Digest of environmental statistics', www.defra.gov.uk/environment/statistics/wastats/archive200611.xls (accessed 1 March 2007).

Defra (2007c) 'Waste strategy for England 2007: incentives for recycling by households', http://www.defra.gov.uk/environment/waste/strategy/incentives/index.htm (accessed 20 July 2007).

Defra (2007d) Household waste and recycling: 1983/4–2005/6, United Kingdom. http://www.defra.gov.uk/environment/statistics/waste/kf/wrkf04.htm

DETR (2000a) 'Consultation paper on implementation of Council Directive 1999/31/C on the landfill of waste. London: DETR. p. 56.

DETR (2000b) 'Waste strategy for England and Wales: Part 2, Norwich: The Stationery Office Limited. pp.10–11.

DETR (2000c) 'Waste strategy for England and Wales: Part 2, Norwich: The Stationery Office Limited. pp.30–1.

Dunnion, K. and Scandrett, E. (2003) 'The campaign for environmental justice in Scotland as a response to poverty in a northern nation' in J. Agyeman, J. D. Bullard and B. Evans (eds), *Just Sustainabilities: Development in an Unequal World*. London: Earthscan.

Environment Agency (2007a) 'Waste overview', www.environment-agency.gov.uk/subjects/waste/1031954/?land=_e (accessed 26 April 2007).

Environment Agency (2007b) 'Environmental facts and figures: landfill', www.environment agency.gov.uk/yourenv/eff/1190084/resources_waste (accessed 23 April 2007).

EU Waste Legislation (2006) Framework waste legislation. Directive 2006/12/EC of the European Parliament and of the Council of 5 April 2006.

Eurostat (2006) *'Municipal waste – kg per person per year'*, http://epp.eurostat.ec.europa.eu/portal/page?_pageid=1996,39140985&_dad=portal&_schema=PORTAL&screen=detailref&language=en&product=STRIND_ENVIRO&root=STRIND_ENVIRO/enviro/en051 (accessed 9 July 2007).

Friends of the Earth (2007). http://www.foe.co.uk/campaigns/waste/issues/reduce_reuse_recycle/index.html

Greenpeace (2005) *Decentralised Energy*. London: Greenpeace.

Greenpeace International (2006) *'Toxic waste in Abidjan: Greenpeace evaluation'*, http://www.greenpeace.org/international/news/ivory-coast-toxic-dumping/toxic-waste-in-abidjan-green (accessed 1 August 2007).

Gregson, N. and Beale, V. (2004) 'Wardrobe matter: the sorting, displacement and circulation of women's clothing', *Geoforum*, 35: 689–700.

Hawken, P., Lovins A., Lovins, L.H. (1999) *Natural Capitalism*. New York: Little Brown and Company.

Health Canada (2006) 'Dioxins and furans', http://www.hc-sc.gc.ca/fn-an/securit/chem-chim/furan/index_dioxin_furan_e.html (accessed 20 April 2007).

HM Treasury (2003) 'Pre-budget report 2002: landfill tax. Norwich: The Stationery Office Ltd.

Holdway, R. (2005) 'Body of evidence', *RSA Journal*, April.

Hore-Lacy, I. (2003) *Nuclear Electricity*. Melbourne, Australia: Uranium Information Centre Ltd/World Nuclear Association. http://www.uic.com.au/ne5.htm (accessed 19 July 2007).

IPPR (Institute for Public Policy Research) and Green Alliance (2006) *A Zero Waste UK*. London: IPPR.

Knox, A. (2005) 'An overview of incineration and EFW technology as applied to the management of municipal solid waste', http://www.oneia.ca/files/EFW%20-%20Knox.pdf (accessed 19 July 2007).

Letsrecycle.com (2005) Letsrecycle/materials/composting/index.jsp (accessed 18 July 2007).

Letsrecycle.com (2007) 'Tyre recycling' http://www.letsrecycle.com/equipment/tyres.jsp (accessed 18 July 2007).

Market Research.com (2006) 'The 2007–2012 world outlook for waste treatment and disposal', www.marketresearch.co./map/prod/1497910 (accessed 30 April 2007).

Matthews, E., Amann, C., Bringezu, S., Fischerkowalski, M., Huttler, W., Kleijn, R., Moriguchi, Y., Ottke, C., Rodenburg, E. and Rugich, D. (2000) 'The weight of nations: material outflows from industrial economies', Washington, DC: World Resource Institute.

Ministry of Housing and Local Government/Scottish Development Department (1970) 'Disposal of solid toxic wastes'. London: HMSO.

Murray, R. (2002) *Zero Waste*. London: Greenpeace Environmental Trust.

Nuclear Information and Resources Service (1992) '"Low level" radioactive waste', www.nirs.org/factsheets/11wfcf.htm (accessed 1 May 2007).

Ollie Recycles (no source date) 'Aylesford newsprint', http://www.aylesford-newsprint.co.uk (accessed 1 May 2007).

Petts, J. (2005) 'Enhancing environmental equity through decision-making: learning from waste management', *Local Environment*, 10(4): 397–410.

Prime Minister's Strategy Unit (2002) 'Waste not: want not'. London: Cabinet Office, p. 61.

Project Integra (2000) 'A history of waste', www.integra.org.uk/facts/history.html (accessed 8 March 2007).

Scottish Executive (2006) 'Environment: national waste strategy', http://www.scotland. gov.uk/Topics/Environment/Waste/17103/8893 (accessed 19 July 2007).

Seyfang, G. (2003) 'Growing cohesive communities, one favour at a time: social exclusion, active citizenship and time banks', *International Journal of Urban and Regional Research*, 39(1): 62–71.

Sills, H. (2007) Project Manager for London Remade, personal communication.

SITA (2004) 'The waste hierarchy – position paper', www.sita.co.uk/assets/PP_WH. pdf (accessed 10 March 2007).

Southern Cross University (2005) 'Radiation safety manual, SCU & Bartolo safety management services', www.scu.edu.au/admin/hr/dds/index.php?action=download&mat_id=1467&site_id=61

The Council of the European Union (1999) 'The Landfill Directive 1999/31/EC', http://europa.eu.int/eur-lex/pri/en/oj/dat/1999/l_182/l_18219990716en00010019.pdf (accessed 19 July 2007).

The Parliamentary Office of Science and Technology (2000) 'Incineration of household waste post 149'. London: The Parliamentary Office of Science and Technology. pp. 1–4.

United Nations (2001) 'Economic and Social Council: Economic Commission for Europe – Committee on Environmental Policy: Regional Consultative Meeting for the World Summit on Sustainable Development', unece.org/env/documents/2001/cep/ac12.2001e.pdf (accessed 30 April 2007).

Walker, G., Fairburn J., Smith G. and Mitchell, G. (2003) *Environmental Quality and Social Deprivation*. Bristol: Environment Agency.

Ward, D.B., Goh, Y.R., Clarkson, P.J., Lee, P.H., Nasserzadeh, V. and Swithenbak, J. (2002) 'A novel energy-efficient process utilizing regenerative burners for the detoxification of fly ash', *Process as-Environ*, 80 (B6): 317–26.

Wasteonline (no source date) 'Legislation affecting waste', http://www.wasteonline.org.uk/resources/InformationSheets/Legislation.htm (accessed 4 March 2007).

Waste Watch (no source date) 'Rethinking waste management to reap rewards: minimising waste for business benefit', http://www.wasteonline.org.uk/resources/WasteWatch/Rethink_WasteManage_ToReap.pdf (accessed 20 July 2007).

Waste Watch (2004) 'Plastics recycling information sheet', www.wasteonline.org.uk/resources/InformationSheets/flashes.htm (accessed 11 April 2007).

Watson, M. and Bulkeley, H. (2005) 'Just waste? municipal waste management and the politics of environmental justice', *Local Environment*, 10(4): 411–26.

Weizsacker, E.V., Lovins, A. and Lovins, L. (1995) *Factor Four: Doubling Wealth, Halving Resource Use*. London: Earthscan.

World Nuclear Association (2007) 'SYNROC', www.world-nuclear.org/info/inf68.html (accessed 29 April 2007).

World Resource Foundation (undated), 'Life cycle analysis and assessment', http://www.gdrc.org/uem/lca/life-cycle.html (accessed 19 July 2007).

8 Global Climate Change

John Woodward and Susan Buckingham

'Warming of the climate is unequivocal, as is now evident from observations of global average air and ocean temperatures, widespread melting of snow and ice and rising global mean sea-level'. (IPCC, 2007)

'The scientific evidence is now overwhelming: climate change presents very serious global risks, and it demands an urgent global response'. (Stern Review, 2006)

Learning outcomes

Knowledge and understanding of:
○ current trends in climate, and atmospheric gas concentrations
○ the extent to which global warming is anthropogenically produced

Critical awareness and evaluation of:
○ the reliability and accuracy of Global Climate Model (GCM) predictions
○ the links between policy responses and the range of interests involved in these
○ the unequal effects of climate change, at different scales, and at different times
○ the relative merits of prevention, mitigation and adaptation of anthropogenic climate change

---

Introduction

In the last 250 years, humans have released ever greater quantities of carbon dioxide (CO_2), methane(CH_4), nitrous oxide (N_2O) and other greenhouse gases, into the Earth's atmosphere. CO_2 has risen from pre-industrial concentrations of 280 ppm to current values in excess of 380 ppm, and is currently rising by 1.9 ppm per year. There is a growing consensus among climate researchers that these greenhouse gases are causing the Earth's temperature to rise. Scientists have measured a temperature rise of 0.76 °C (with confidence intervals of 0.56 to 0.92 °C) between 1850 and 2005, as a result of increased radiative forcing from the increases in atmospheric greenhouse gases. (IPCC, 2007). Analysis of deep ice cores from Antarctica and Greenland reveal that the rise of atmospheric greenhouse gas concentrations, and the associated warming this generates, is occurring at a rate faster than any natural change seen in the last 650,000 years.

Despite such evidence, the wide-ranging and complex issues embedded in the climate change debate have vexed scientists, business representatives, environmentalists and politicians. Judgements and agendas permeate deeply controversial interest group biases and political approaches (O'Riordan and Jäger, 1996). This is, in part, due to the fact that, despite the annual cost of current climate change research being estimated at over three billion dollars (Stanhill, 2001), there are still large gaps in our understanding of the natural system and large inherent uncertainties in our predictions (Shackley et al., 1998). Indeed, the recent report from the Intergovernmental Panel on Climate Change (IPCC, 2007) excludes an estimate of sea-level rise produced by the melting of the ice sheets of Greenland and Antarctica, due to the current uncertainties within these systems. It is also hard to summarise complex climate research into easily digestible snippets for politicians and policymakers. Furthermore, the information available on climate change for policy-making purposes is plagued by large inherent uncertainties (Shackley et al., 1998).

This chapter will show how complex and uncertain climate change is, and some of the reasons behind this, including the methods, errors and outcomes associated with modelling the global climate. Prior to 2001, this complexity and uncertainty served to keep climate change low on the global political agenda, as different interest groups used this to avoid effective legislation. The current climate of political engagement is patchy at the global scale, as the chapter will show. O'Riordan (2004) argues that the 'target' of climate change is too nebulous to galvanise public interest, and renders people susceptible to negative 'sacrifice' discourses put forward by government and business. Other factors, as this chapter will also show, include the time and space dissonances which mean that the causes of global climate change are often distant from their effects. Commonly, the people or communities involved in producing the causes are not the same as those most affected by them with, arguably, the worst effects felt by communities least able to deal with them, a discussion which is also developed in Chapter 9 on natural hazards and elsewhere in the book.

The science of global warming will then be integrated into the policy interface by considering the role of the Intergovernmental Panel on Climate Change (IPCC) and other international institutions, which have largely set a global policy agenda to which individual states have had to respond. Moreover, strong political and economic interests hold powerful sway over national and international legislative bodies and affect their capacity to construct effective and just legislation.

Another factor making it difficult to gain a clear understanding of global climate change is that there is a lack of agreement on its importance as a critical global problem. While the European Union claims that global warming is the most serious international problem (eclipsing even terrorism in this era of post 9/11), to which the introduction to the book has already made reference, it is not only North Americans and Australians who disagree. Politicians and NGOs in the Global South challenge

Table 8.1   Greenhouse gases basketed under the Kyoto Protocol and their main generators (Note: greenhouse gases produced by air transport are exempt from the Protocol.)

Greenhouse Gas	Main Sources
Carbon dioxide (CO_2)	Fossil fuel combustion (e.g. road transport, energy industries, other industries, residential, commercial and public sector); forest clearing
Methane (CH_4)	Agriculture, landfill, gas leakage, coal mines
Nitrous oxide (N_2O)	Agriculture, industrial processes, road transport, other
Perfluorocarbons (PFCs)	Industry (e.g. aluminium production, semi-conductor industry)
Hydrofluorocarbons (HFCs)	Refrigeration gases; industry (as perfluorocarbons)
Sulphur hexafluoride (SF_6)	Electrical transmissions and distribution systems, circuit breakers, magnesium production

Source: UNFCCC, 2003

this for very different reasons, citing the immediate problems facing their countries, whether this be HIV/AIDS, malaria, famine and malnutrition, lack of access to clean drinking water or effective sewage disposal. Such debates were fuelled at the turn of the 21st century by a book, *The Skeptical Environmentalist,* in which a Danish statistician, Bjorn Lomborg, argued at length that data forecasting global climate change is flawed (Lomborg, 2001). Unsurprisingly, given the ability of data to be forever manipulated, Lomborg's arguments and the data on which he bases them are also hotly debated. His more recent ranking exercise of global problems through the 'Copenhagen Consensus', and other controversies sparked by climate change sceptics, will be addressed later in the chapter. This chapter will conclude with possible responses to global warming, including issues of mitigation and/or adaptation, drawing on the Stern Review, commissioned by the UK Treasury to investigate the economics of climate change, but which has had far-reaching international effects.

While this chapter does not dwell on the anthropogenic causes of climate change, Table 8.1 offers a brief summary.

What is global warming?

Life on Earth is supported by a bubble of gas some 30 km thick, held in place by the gravitational pull of our planet. This atmosphere provides the oxygen we need to breathe and to protect us from the extremes of cold and some parts of the electromagnetic spectrum output by the sun, for example UV light, which is partly absorbed by ozone. Without the protection of the atmosphere, the surface of the Earth would be some 33 °C colder. Greenhouse gases, such as water vapour (H_2O), carbon dioxide (CO_2), methane (CH_4), chlorofluorocarbons (CFC) and nitrous oxide (N_2O) trap heat in the atmosphere. In a stable climate, the concentration of greenhouse gases in the atmosphere will remain reasonably constant and play an important role in the Earth's energy balance (Figure 8.1a). Figure 8.1b illustrates the greenhouse effect.

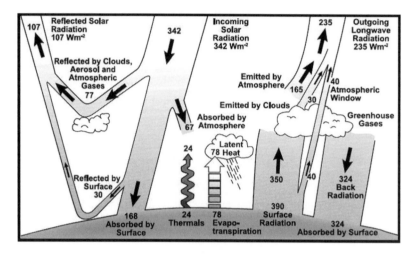

Figure 8.1a Estimate of the Earth's annual and global mean energy balance. Over the long term, the amount of incoming solar radiation absorbed by the Earth and atmosphere is balanced by the Earth and atmosphere releasing the same amount of outgoing longwave radiation. About half of the incoming solar radiation is absorbed by Earth's surface. This energy is transferred to the atmosphere by warming the air in contact with the surface (thermals), by evapotranspiration and by longwave radiation that absorbed by clouds and greenhouse gases. The atmosphere in turn radiates longwave energy back to Earth as well as out to space.

Source: Kiehl and Trenberth, 1997 in IPCC, 2007

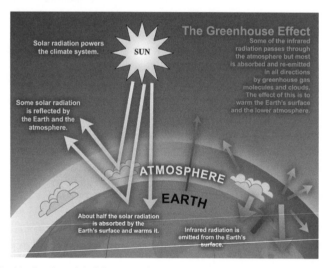

Figure 8.1b An idealised model of the natural greenhouse effect.

Source: IPCC, 2007

The link between increased concentrations of CO_2 in the atmosphere and climate change is only now becoming apparent at a global scale. Most scientists believe that proof exists of a steady warming, which can be linked to the increase in greenhouse gas concentrations, though the exact impact of such changes remains open to speculation.

Box 8.1   Scientific definitions

Climate change: any change in the global climate system, over time, whether due to natural variability or as a result of human activity.

Global warming: recent climatic amelioration, observed across the globe, believed to be a result of anthropogenic (human) forcing due to the increased release of greenhouse gases into the atmosphere.

Greenhouse gas: A gas that absorbs and emits radiative energy. The effects of the blanket of greenhouse gases making up our atmosphere are to trap heat at the Earth's surface, warming the Earth by some 33 °C.

Electromagnetic radiation: Energy generated by the interaction of electric and magnetic fields. Energy arrives at the outer edge of the Earth's atmosphere as short-wave radiation, centred in the visible part of the radiative spectrum. Energy leaves the Earth, primarily as long-wave radiation centred on the infrared band of the spectrum, though some long-wave light is reflected straight back to space. Greenhouse gases absorb long-wave radiation, re-radiating some of this back towards the Earth's surface, thereby warming the Earth.

Radiative forcing: a measure (in Watts per square metre (Wm^{-2})) of the influence of a factor that alters the balance of incoming to outgoing radiation from the earth. Greenhouse gases are a positive forcing, as they cause the Earth to warm.

El Niño southern oscillation: (ENSO or El Niño) is a natural variation in the Pacific ocean-atmosphere system. The coast of Peru is normally an up-welling region of cold, deep ocean water, which replaces surface water driven westward by the trade winds. During ENSO events, every 3–7 years, the trade winds weaken, reducing up-welling, resulting in warmer ocean temperatures and cloudiness across the central and eastern Pacific. ENSO is thought to result in: fewer Atlantic hurricanes; droughts in Brazil, Australia, Africa and Indonesia; and heavy rainfall on the arid coast of South America.

Observed changes in greenhouse gas concentrations and climate

Trends in greenhouse gas concentrations

Greenhouse gas concentrations in the atmosphere have been rising steadily since the beginning of human cultivation of crops in Asia, some 8000 years ago. This concentration has risen ever more rapidly since the beginning of the Industrial Revolution around 1750. For the last 150 years, earth scientists have been measuring the concentrations of greenhouse gases, and have established that the volume of CO_2 in the atmosphere has increased by more than a third since 1750. The present CO_2 concentration has not been exceeded during the past 420,000 years, while the present rate of increase is unprecedented during the last 20,000 years. Almost all this rise is attributable to anthropogenic emissions of CO_2, with 75 per cent thought to be due to fossil fuel burning, and the rest predominantly due to land use change, such as deforestation (IPCC, 2001).

CO$_2$ levels in the atmosphere, measured at Mauna Loa in Hawaii since 1957, show the recent steady rise in atmospheric CO$_2$ (Figure 8.2). Pre-industrial (1750) atmospheric concentrations were around 280 ppm, with 2005 values at 379 ppm (i.e. 379 molecules of CO$_2$ gas per million molecules of dry air). Current annual emissions, driving this rapid rise in CO$_2$ concentrations, are estimated to be in the region of 7.2 GtC (gigatonnes of carbon). The Mauna Loa curve shows that in 2001–2002 and

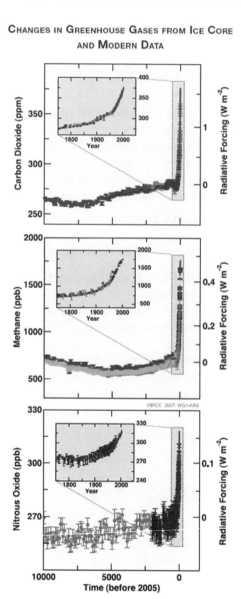

Figure 8.2 Atmospheric concentrations of carbon dioxide, methane and nitrous oxide over the last 10,000 years (large panels) and since 1750 (inset panels). Measurements are shown from ice cores (symbols with different shades of grey for different studies) and atmospheric samples. The corresponding radiative forcings are shown on the right-hand axes of the large panels
Source: IPCC, 2007

2002–2003, the concentrations of CO_2 rose by 2.08 ppm and 2.54 ppm respectively. This was the first recorded instance of a rise of more than 2 ppm for two consecutive years. Under normal conditions, significant rises in CO_2 levels in Hawaii are associated with El Niño years. The recent trend suggests that these patterns may be being overprinted by the signal from anthropogenic CO_2 emissions.

CO_2 is not the only important greenhouse gas to have seen rapid post-industrialisation rises in atmospheric concentration. Methane has increased from a pre-industrial value of around 715 ppb to a 2005 value of 1774 ppb. The IPCC (2007) states that this rise is predominantly due to agriculture and fossil fuel use, though sources are not well constrained and there is a suggestion in the recent record that emission levels may have stabilised during the 1990s. Nitrous oxide, predominantly released due to intensive agriculture, has also risen from 270 to 319 ppb between 1750 and 2005. These increased emissions have led to the IPCC (2007) stating that there is 'a very high confidence' that radiative forcing has risen by +1.6 (within the error bounds +0.6 to +2.4) Wm^{-2}, due to anthropogenic influences on the atmospheric system.

Trends in temperature

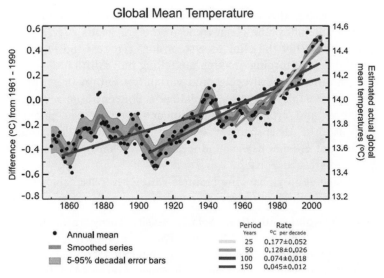

Figure 8.3 Annual global mean observed temperatures (black dots) with linear fits to the data. The left hand axis shows anomalies relative to the 1961 to 1990 average and the right hand axis shows the estimated actual temperatures both in °C. Linear trends for the last 25, 50, 100 and 150 years are shown as straight lines, and correspond to 1981 to 2005, 1956 to 2005, 1906 to 2005, and 1856 to 2005, respectively. Note that for shorter recent periods, the slope is greater, indicating accelerated warming. The thick grey curve is a smoothed series depiction to capture the decadal variations. To give an idea of whether the fluctuations are meaningful, decadal 5% to 95% (light grey band) error ranges about that line are given.

Source: IPCC , 2007

The general temperature curve seen in Figure 8.3 is made up of observational data showing changing global average temperatures over the last 150 years. Four stages can be identified within the overall trend:

1   A long, very irregular but generally cool period between 1860 and 1910.
2   A very rapid, regular and prolonged period of warming between 1910 and 1943.
3   An equally long period of small and irregular cooling between 1943 and 1975.
4   A rapid warming phase since 1975. Eleven of the twelve years between 1995 and 2006 rank among the 12 warmest years in the global instrumental temperature record which dates from 1850.

The overall trend suggests an average global temperature increase between the periods 1850–1899 and 2001–2005 of 0.76 °C (with confidence intervals 0.57 to 0.95 °C) (IPCC, 2007). Varying degrees of warming are recorded for every continent (with the exception of Antarctica where the instrumental record is inconclusive due to limited meteorological data), both on land and over the oceans. Temperature rises are also seen throughout the lower and mid-troposphere, the 11 km of the Earth's atmosphere nearest the surface.

What else might control the temperature curve?

If there were a direct link between greenhouse gas emissions and temperature, then the emissions and temperature curves would follow the same trend. Figure 8.3, showing the temperature record, while describing an overall increase, does not follow the clean trend of the emissions curves (Figure 8.2). This implies that either the climate system does not respond in a linear fashion to greenhouse gas forcing (making modelling more difficult – see below) or that there are other, natural cycles and variability within the climate system controlling the Earth's response to changing concentrations of greenhouse gases. Scientists and pressure groups, unconvinced by the global warming arguments, use the discrepancies between these curves to argue against the importance of greenhouse gas-induced climate change. So what else might control the climate's response?

Significant change in global temperatures can, with 'reasonable scientific certainty', be linked to three other causal mechanisms (see Houghton, 2004):

1   **Changes in solar radiation**: Solar intensity varies through the 11-year sunspot cycle by about 0.2–0.5 Wm^{-2}. There are known historic variations, such as the Maunder Minimum between 1650 and 1700 when there were no sunspots. This was the coldest period in Europe for the last 1000 years, and was the core of the Little Ice Age. Solar intensity can, therefore, control our climate, and produce significant temperature variations at short timescales. There is significant uncertainty in scientific measurements of sunspot activity and solar intensity before the 1970s, so there are only very limited usable historical data-sets. The proxy data indicating sunspot activity can explain over 60 per cent of climate variance prior to the 1970s. Recent research indicates that the observed rapid rise in global mean temperatures post-1985 cannot be attributed to solar forcing (Lockwood and Fröhlich, 2007).
2   **Volcanic eruptions**: The injection of sulphate aerosols and dust into the atmosphere during large volcanic eruptions will cool the climate. The 1991 eruption of Mount Pinatubo cooled the planet by 0.5 °C for several years. Large eruptions can cause a negative forcing of up to 3–5 Wm^{-2}. In extreme cases, scientists estimate this may be as large as 12 Wm^{-2}.

3   **Release of aerosols:** Anthropogenic sulphate aerosols are known to have a negative forcing on the climate system. The effect is also localised, so is often seen downwind of conurbations. The cooling forcing has risen from the early 1900s to account for around -1 Wm^{-2}.

This suggests that CO_2 is significantly more important as a mechanism than short-term sunspot/solar radiation forcing, volcanic eruptions or aerosols. The problem for climate modellers and their attempted predictions of climate change is that the climate system is far more unpredictable than future CO_2 emissions trends.

Examples of change

Glaciers

The widespread 20th-century retreat of valley mountain glaciers in non-polar regions is well documented and is estimated to contribute 0.2 to 0.4 mm yr^{-1} of sea-level change. Any significant future rises in sea level will be determined by changes in the mass of the polar ice sheets in Antarctica and Greenland, which together hold some 33 million km^3 of ice, enough to raise global sea level by some 70 metres. The state (and fate) of these large ice sheets is less well understood than their valley glacier cousins. This is complicated, as the ice sheets are not only changing as a result of global warming, they are also undergoing long-term adjustments and are losing mass as a result of natural climate change since the last glacial maximum, 20,000 years ago. Box 8.2 presents a case study identifying some of the scientific uncertainties of dealing with risk, regarding the break-up of the West Antarctic Ice Sheet (WAIS).

Box 8.2   The complexity of predicting the behaviour of the West Antarctic Ice Sheet

The relationship between scientists and policymakers is often perceived as contentious. In part, this is because scientists work on detailed reports with discussions of errors and probability associated with any predicted impact or trend. For policymakers and the public, this must be condensed into usable sound bites. This example will show an attempt to provide policymakers with usable information on the possible future collapse of the West Antarctic Ice Sheet.

The Antarctic Ice Sheet stores 25.4 million km^3 of ice, equating to 90 per cent of the freshwater on Earth. The future of the ice sheet under climate change scenarios is unknown. This is because there are large uncertainties in accumulation (the addition of mass, primarily through snowfall) and ablation (the loss of mass, predominantly through iceberg calving) measurements, particularly as there is likely to be an increase in precipitation due to warming of the Southern Ocean.

Antarctica is made up of two large ice sheets, the East Antarctic Ice Sheet, grounded on bedrock above sea level, and thought to be stable and predictable in its response to climate change, and the West Antarctic Ice Sheet, grounded below sea level, and termed a marine ice sheet. Because of the marine base of WAIS, scientists have

(Continued)

expressed concerns regarding WAIS and predicted possible rapid collapse of the ice sheet. In the media, this is often simplistically associated with the disintegration of ice shelves, such as the Wordie (see Figure 8.4) and Larsen Ice Shelves, in the Antarctic peninsula.

Figure 8.4   Icebergs in Wordie Bay produced by the break-up of the Wordie Ice Shelf on the Antarctic Peninsular

Source: author

Vaughan and Spouge (2002) define collapse as sea-level rise of 1 m per century or 4 m total. Significant sea-level rise is 0.2 m per century or >1 m in total. Vaughan and Spouge (2002) document three methods for estimating the risk of WAIS collapse:

1 Status of past behaviour of WAIS: WAIS has collapsed before but the evolution of WAIS is not well recorded, and atmospheric conditions are different today compared to those during the last collapse.
2 Risk estimate based upon model predictions: computer models of WAIS suggest instability but often fail to agree on the timing of any collapse.
3 Risk estimate using Delphi technique: this encapsulates the current state of expert opinion through a series of questions – respondents use model results, literature or gut feeling to make value judgements about risk.

Vaughan and Spouge conclude that the experts do not agree about when or whether WAIS will collapse. They find considerable benefit in the Delphi technique, as this methodology is understood by policymakers who often apply the technique of quantitative risk estimation and risk assessment. The process helps to rank particular issues in a debate and provides a measure of the degree of accord within the scientific community.

After questioning scientists, they estimate a 5 per cent probability of collapse and 30 per cent probability of significant sea-level rises in the next 200 years. This produces easily digestible information for policymakers. The scientists have helped digest the complex facts into a simple statement of possible risk. Policymakers can then decide where this risk fits within their assessment of environmental concerns.

Hurricanes

Media coverage often invokes large hurricanes, such as Katrina, as proof of global warming. Recent research, summarised in Shepherd and Knutson (2007), suggests that there has been a recent increase in both the intensity and activity of Atlantic hurricanes, mirrored by increases in tropical cyclone activity. As in other areas of earth system science, however, the problem is assessing and proving the link between hurricane activity and anthropogenic-induced global warming (the link between cause and effect). In order to generate a hurricane, six key atmospheric factors must usually be met: warm ocean waters (higher than 26.5 °C) through a sufficient depth of water; unstable atmospheric conditions; a moist mid-troposphere; sufficient distance from the equator to generate cyclonic behaviour; some form of near-surface disturbance to help generate the vortex, such as tropical easterly waves; and low values of vertical wind shear (changing wind speed with height). Which, if any of these processes are being significantly altered by climate change? The most plausible predictions indicate that any change in intensity and activity, as a response to warmer ocean temperatures, is likely to be locally variable, in response to changes in the larger atmospheric system. As Shepherd and Knutson (2007) conclude, 'significantly more research – from observations, theory and modelling – is needed to resolve the current debate around global warming and hurricanes'.

**Box 8.3   Summary of IPCC observations of recent
climate change**

The 2007 IPCC report summarises the key scientific observations on climate change. It is impossible to replicate all observations in this chapter, so a summary of some key findings is given below.

There is high confidence that effects have been documented in the following systems:

- Snow, ice and frozen ground (including permafrost) with change exemplified by: retreating glaciers and Arctic sea ice; enlargement and an increased number of glacial lakes; increasing ground instability in permafrost regions; increasing rock avalanches in mountain regions; and increased run-off and earlier spring peak discharge in many glacier- and snow-fed rivers.
- Terrestrial biological systems, including such changes as: earlier timing of spring events, such as leaf-unfolding, resulting in earlier greening of vegetation; earlier bird migration and egg-laying; changes in some Arctic and Antarctic ecosystems, including those in sea-ice biomes, with predators high in the food chain (such as polar bears) being affected around the world; pole-ward and up-ward shifts in ranges for plant and animal species, often linked to longer thermal growing seasons.
- Marine and freshwater biological systems are associated with rising water temperatures, as well as related changes in ice cover, salinity, oxygen levels and circulation. Effects include: shifts in ranges and changes in algal, plankton and fish abundance in high-latitude oceans; increases in algal and zooplankton abundance in high-latitude and high-altitude lakes; range changes and earlier migrations of fish in rivers.

(Continued)

There is a medium level of confidence that effects of temperature increases have been documented in the following:

- Agricultural and forestry management at northern hemisphere higher latitudes, resulting in earlier spring planting of crops, and alterations in disturbance regimes of forests due to fires and pests.
- Aspects of human health, such as heat-related mortality in Europe, infectious disease vectors in some areas, and allergenic pollen in northern hemisphere high and mid-latitudes.
- Some human activities in the Arctic (e.g. hunting and travel over snow and ice) and in lower elevation alpine areas (such as mountain sports).

Further, there are indications that:

- settlements in mountain regions are at enhanced risk to glacier lake outburst floods caused by melting glaciers
- in the Sahelian region of Africa, warmer and drier conditions have led to a reduced length of growing season with detrimental effects on crops
- in southern Africa, longer dry seasons and more uncertain rainfall are prompting adaptation measures
- sea-level rise and human development are together contributing to losses of coastal wetlands and mangroves and increasing damage from coastal flooding in many areas.

Global Climate Models (GCMs)

What are GCMs?

The complexity of the climate system and the absence of definitive analogues for our evolving climate have led to the use of theoretical computer models known as global climate models (GCMs) to attempt to predict the future influence of greenhouse gases (MacCracken et al., 1991). GCMs predict climate based upon our understanding of the physical laws controlling the environment. These models represent what are believed to be the most important aspects of the atmosphere, ocean, cryosphere and biosphere, though they cannot yet simulate all aspects of the climate (IPCC, 2001).

GCMs divide the surface of the Earth into cells, usually some 100 km by 100 km in size. The resolution is important as, ideally, cell size would mirror land use on the ground surface. Clearly, this is not the case, as cell resolution is closely controlled by computer capability. Extra cells result in longer model runs and increase the number of links between cells, making the modelling process more complex. Once each cell area has been defined as a grid, a series of vertical layers are added to the model, creating boxes that represent the atmosphere. Near the ground surface, layers are small, often in the order of metres. This allows the boundary layer, the link between the ground surface and the atmosphere, to be modelled in significant detail. This is where the most important climatic interchanges take place, controlling moisture and heat transfer from the Earth's surface to the lower atmosphere. As elevation increases, layer depths become larger, as the relative importance of interactions decreases, and our ability to measure parameters as input to the models also decreases. GCMs then require each column of cells to be linked to the cells around them, so that transfer for heat and moisture can be modelled.

Such interactions are extremely complex, and must be carried out at every time interval for which the model is being run. Due to the significant changes in meteorological conditions throughout the day, it is common to run GCMs at hourly steps. This daily change is largely controlled by the primary input required to initiate a GCM, namely incoming solar radiation. This and other key controls are entered into the model, such as atmospheric gas concentrations and factors associated with the basic physics of the Earth, such as oceanic and atmospheric circulation patterns. The model then calculates for every vertical box, for every grid cell, and for every time-scale, the key meteorological parameters associated with wind speed, precipitation and temperature and so on. These calculations are parameterised (see Box 8.4 for an explanation of this term) against current known climatic conditions and variables. The next problem to be faced by modellers is what will happen in the future. Neither the Earth system nor the population that lives in it is static, so this dynamism must be factored into model scenarios.

Box 8.4   How big is a cloud?

Clouds are an important component of the atmosphere, as they have a direct effect on the temperature and moisture profile of the atmosphere and control the amount of incoming solar radiation that reaches the surface of the Earth. When including clouds in climate models, it is important to know the elevation and thickness of the cloud (low clouds tend to cool the system, while high clouds warm the atmosphere), the cloud type (the presence of water or ice in a cloud will control the cloud's optical properties) and, because clouds are forming at sub-grid scale, the percentage cover of clouds (individual clouds cannot be modelled by GCM) (Houghton, 2004). Small errors in calculating any of these controls have a disproportionate effect on the accuracy of a GCM. Unfortunately, many models make use of simplified parameters (numerical constants that constrain physical equations) when dealing with clouds:

'Many physical processes, such as those related to clouds, take place on much smaller time scales and therefore cannot be properly resolved and modelled explicitly, but their average effects must be included in a simple way by taking advantage of physically based relationships with the larger scale variables (a technique known as parameterization).' (IPCC, 1996)

Thus, it seems likely that important processes are often absent or highly simplified in ways that might affect model sensitivity. Models with different cloud parameterisation dramatically affect estimated sensitivity (MacCracken et al., 1991). Validation is weakened by poor observational data-sets of cloud feedback effects. The correct representation of clouds within GCM remains one of many difficult challenges facing climate scientists.

Emissions scenarios – the basis for predicting future climate change

In order to use a computer model to predict future climate change, modellers require clear statements regarding the future emissions of greenhouse gases expected over the coming century. The emissions of greenhouse gases are likely to depend upon complex economic and political issues, such as land use change, industrial development and global population dynamics. Since 1996, the IPCC has been developing new emission scenarios following four different narrative storylines. These are:

A1:   Very rapid economic growth, a peak in global population by the mid-21st century, rapid development of more efficient technologies, with convergence between developed and developing regions. The A1 scenario is further subdivided into fossil intensive (A1F1), non-fossil energy sources (A1T), or a balance across sources (A1B).

A2:   A very heterogeneous world, with preservation of local identities. Economic development is locally focused, resulting in slower development but continuous population increases.

B1:   A convergent world economy, similar to scenario A1, though with a heavy focus on clean and resource-efficient technologies, with an emphasis on global solutions to economic, social and environmental sustainability.

B2:   Again, a heterogeneous world similar to A2, though with an emphasis on local solutions to economic development, social and environmental sustainability.

The combination of factors related to the four narratives result in 40 scenarios for the GCMs to model climate change. GCM outputs from different modelling groups can then be compared for each scenario. There is as much complexity in the development of the scenarios as there is in the models themselves.

How are the models tested?

Models can be tested against their ability to predict:

- The current climate – models are run against control data (meteorological records) in order to assess how well they reconstruct current climate. Parameterisation (the application of numerical constants to constrain physical equations) is applied to the model to increase its ability to predict current climate. Care must be taken to apply this parameterisation within the constraints of our understanding of the physical processes active within the environment. Climate change sceptics frequently cite model parameterisation as a key problem with GCM predictions, referring to this process as a 'fudge-factor'. Despite this, the current state-of-the-science GCMs, using well constrained equations, successfully recreate latitudinal and seasonal changes in climate.

- Recent climate change – models must be run for considerable time periods to reach steady-state conditions, that is to recreate current climate. Model results can be compared to known temperature trends for the last century, and are able to recreate small climate changes, such as short periods of cooling due to volcanic activity, for example, the cooling effect of the eruption of Mount Pinatubo in 1991 (Houghton, 2004).

- Past Earth climates – models can also roughly explain (within the constraints of geological evidence) the changes in Earth's temperature throughout its geological history, as solar intensity has increased and atmospheric composition has changed (MacCracken et al., 1991).
- The climates of other planets in the solar system with known atmospheric conditions – using the same physical assumptions as applied to the Earth, models have successfully predicted the climates of other planets in the solar system, such as Venus and Mars.

The success of models to recreate current climate (short time-scales) and past climates (long time-scales) suggests that the basic physics applied in the models are reasonably well understood. Unfortunately, changes in greenhouse gases dominate the decadal-to-century time-scales and there are few appropriate tests for this (mid) time-scale (MacCracken et al., 1991). This limits our confidence in the ability of GCMs.

What limits confidence in the models?

'The limitations of computer modelling, the unrealistic nature of the basic assumptions made about future technological change and political value judgements have often distorted the scenarios presented to the public' (Lomborg, 2001).

Different models with differing parameterisation linked to 40 scenarios from four different narratives for predicting future change provides an extremely complex and challenging environment for climate change modellers, leaving many sceptical as to the usefulness of GCM predictions. In order to be convinced that the predictions provided by the models are believable, scientists must be satisfied that the models are comprehensive and correct. In order to make the models more acceptable and useful, they must be related to the scale of the process involved (see Box 8.4); represent sections of the Earth/atmosphere in as accurate a manner as possible; be as simple as possible; and have as few parameters as possible. MacCracken et al. (1991) suggest our confidence in the models is limited by a number of issues:

1 Many processes are not represented in the models. In part, this is a result of scale, with many meteorological phenomena and surface conditions being below the grid scale of the models. This is also a result of the incremental development of increasingly sophisticated GCMs. In the late 1990s, models failed to adequately incorporate the carbon cycle and non-sulphate aerosol calculations. It was only in the early 2000s that dynamic vegetation and coupled atmospheric chemistry components were added to GCMs.
2 Models are tuned (forced) to achieve improved agreement with observed climate.
3 There are limitations in observational data-sets with which to test the models, for example, the majority of meteorological records are for coastal areas, rather than the interior of the large continents, so we have only limited control over large spatial areas of the models. The old computer adage 'garbage in, garbage out' is only too true of GCM modelling.

4   The models assume that climate will change gradually (rather than in fits and starts) and also that the system is in equilibrium. In reality, the climate system is chaotic and inherently unpredictable at the small scale. An example of this is the problem of modelling the El Niño southern oscillations (ENSO), a chaotic and statistically random circulation phenomenon in the South Pacific. While Houghton (2004) states that some models are now capable of simulating some aspects of ENSO, many aspects are not well simulated in GCMs. If we don't fully understand ENSO, how do we force it to happen in the future, when we do not know how or if it will be generated?

5   There is insufficient data to adequately and simultaneously model non-greenhouse gas forcing (including solar variation, aerosol injections and oceanic circulation changes) in GCMs.

6   Model predictions do not always agree with observations. The weighting towards the influence of greenhouse gases results in models suggesting a more rapid and larger warming than has been observed to date, and irregularities such as the cooling events in the 1970s are not well modelled.

Current GCMs are become ever more sophisticated. Models are able to integrate more and more elements of the Earth's climate system and represent complexities in a more realistic way, thereby generating results comparable with expanding observational data-sets. Comparison between models is helping identify strengths and weaknesses, while nesting of models that run at different resolutions (such as local hydrological models coupled to GCMs) is assisting with problems associated with the scale of many processes within the climate system. Despite these huge advances in state-of-the-science computers and models, the use of computer models and the results they generate is not, and probably will never be, as conclusive as we would like.

Model predictions of future change in greenhouse gas concentrations and climate

February 2007 saw the agreement of the fourth assessment report of the IPCC. This report summarises the most likely future changes in climate as predicted by the range of greenhouse gas emission scenarios presented above. The following subsections highlight some key conclusions of the fourth IPCC report, while Box 8.5 reviews some of the results of climate change predicted by the IPCC.

Box 8.5   IPCC predictions for future change as a result of climate change

The IPCC report (2007) summarises the key scientific predictions resulting from climate change. Some of these observations are summarised below:

- **Fresh water resources and their management:** by mid-century, annual average river run-off and water availability are projected to increase by 10–40% at high latitudes and in some wet tropical areas, and decrease by 10–30% over some dry regions at mid-latitudes and in the dry tropics. Drought-affected areas will likely increase in extent. Heavy precipitation events, which are very likely to increase in frequency, will augment flood risk. In the course of the century, water supplies stored in glaciers and snow cover are projected to decline, reducing water availability to one-sixth of the world's population.

- **Ecosystems:** the resilience of many ecosystems is likely to be exceeded this century by an unprecedented combination of climate change, associated disturbances (e.g. flooding, drought, wildfire, insects, ocean acidification), and other global change drivers (e.g. land use change, pollution, over-exploitation of resources). Approximately 20–30% of plant and animal species are likely to be at increased risk of extinction if the rise in global average temperature exceeds 1.5–2.5 °C. The progressive acidification of oceans due to increasing atmospheric carbon dioxide is expected to have negative impacts on marine shell-forming organisms, such as corals, and their dependent species.

- **Food and forest products:** crop productivity is projected to increase slightly at mid-to high latitudes though will decrease at lower latitudes, especially in seasonally dry and tropical regions. Globally, the potential for food production is projected to increase with rises in local average temperature over a range of 1–3 °C, but above this it is projected to decrease. Increases in the frequency of droughts and floods are projected to affect local production negatively, especially in subsistence sectors at low latitudes. Globally, commercial timber productivity will rise modestly with climate change in the short- to medium-term, though there will be large regional variability around the global trend.

- **Coastal systems and low-lying areas:** coastal systems are likely to be exposed to increasing risks, including coastal erosion, due to climate change and sea-level rise. Increasing human-induced pressures on coastal areas will exacerbate the effect. Coastal wetlands, including salt marshes and mangroves, will be negatively affected by sea-level rise. Many millions more people are projected to be flooded every year, due to sea-level rise by the 2080s. Those densely populated and low-lying areas where adaptive capacity is relatively low, and which already face other challenges, such as tropical storms or local coastal subsidence, are especially at risk. The numbers affected will be largest in the mega-deltas of Asia and Africa while small islands are especially vulnerable.

- **Industry, settlement and society:** costs and benefits for this sector will vary widely by location and scale. The most vulnerable industries, settlements and societies are generally those in coastal and river flood plains, those whose economies are closely linked with climate-sensitive resources; and those in areas prone to extreme weather events, especially where rapid urbanisation is occurring. Poor communities can be especially vulnerable, in particular those concentrated in high-risk areas with limited adaptive capacities (see also Chapter 9 on natural hazards). Where extreme weather events become more intense and/or more frequent, the economic and social costs of those events will increase.

- **Health:** projected climate change-related exposures are likely to affect the health status of millions of people, particularly those with low adaptive capacity. This will occur through: increases in malnutrition and consequent disorders, with implications for child growth and development; increased deaths, disease and injury due to heat waves, floods, storms, fires and droughts; the increased burden of diarrhoeal disease; the increased frequency of cardio-respiratory diseases due to higher concentrations of ground-level ozone related to climate change; and the altered spatial distribution of some infectious disease vectors.

Source: IPCC, 2007

Future changes in temperature

For the next two decades, a warming of 0.2 °C per decade is projected. Even if the concentrations of greenhouse gases and aerosols stabilised at 2000 values, a rise in temperature of 0.1 °C per decade would be expected. Over the longer term, projected, globally averaged, surface warming for the end of the 21st century (2090–2099) relative to 1980–1999 suggests a best estimate for the low scenario (B1) (see p. 188 for scenario descriptions) is a temperature rise of 1.8 °C (likely range 1.1 °C to 2.9 °C) and a worst estimate for the high scenario (A1F1) is a warming of 4.0 °C (likely range 2.4 °C to 5.8 °C). Warming is predicted to be greatest over land and at most high northern latitudes, and least over the Southern Ocean and parts of the North Atlantic. Snow cover is projected to contract, and sea-ice is likely to shrink in both polar regions, with Arctic sea-ice almost disappearing by late summer by the latter part of the 21st century. There is also predicted to be an increase in hot extremes, i.e. heatwaves.

Future changes in precipitation and storminess

It is considered very likely that precipitation levels will increase at high latitudes while decreasing at lower subtropical regions (possibly by as much as 20 per cent in the A1B scenario). Heavy precipitation events will continue to become more frequent, and extra-tropical storm tracks are projected to move pole-ward, resulting in changes in wind, precipitation and temperature patterns. It is likely that tropical cyclones (typhoons and hurricanes) will become more intense, with stronger winds and more heavy precipitation, though there is a suggestion that they might decrease in number.

Predictions for sea-level rise

Over half the world's population live within 60 km of the coast (Holligan and deBoois, 1993). High-density populations on deltaic areas of the world are found in China, Bangladesh and Egypt, making them particularly susceptible to sea-level rise (Nicholls and Leatherman, 1995). Bangladesh, Senegal, Nigeria and Egypt appear particularly vulnerable – that is, they have the least ability to cope with sea-level rise, based on their existing physical and human susceptibility: large and expanding coastal populations and limited experience in adaptation techniques (see Chapter 9). Countries such as Senegal and Uruguay are among many countries threatened by sea-level rise for a very different reason – their economies are highly dependent on beach tourism (Nicholls and Leatherman, 1995). However, it is low-lying, small island developing states (SIDS) that are the greatest cause for concern, and this has been a contributory factor in a United Nations focus on SIDS in the World Summit on Sustainable Development Plan of Implementation (United Nations, 2002: Section VII).

Sea level is estimated to have risen at a rate of 1.8 ± 0.1 mm yr^{-1} over the last century (Douglas, 1991; IPCC, 2001). Estimates predict sea level will continue to rise at this rate, though may increase threefold, with total sea-level rise in 2090–2099 relative to 1980–1999 predicted to be 0.18–0.59 metres (IPCC, 2007). Table 8.2 sets out the main variables involved in sea-level change. These include:

- Thermal expansion of the oceans due to warming sea-surface temperatures – as the near-surface ocean water warms, the density of the water decreases, resulting in thermal expansion. Over decadal and millennial timescales, deep ocean water will also warm, as a result of thermal transfer and mixing due to ocean currents. Thus, even if temperature increases were stopped today, thermal expansion would continue far into the future.
- Change in the terrestrial storage of water – estimates of this reservoir from 1910–1990 are in the range –1.1 to 0.4 mm yr^{-1} of sea-level rise. Such discrepancy indicates the complexity of estimating this contribution, which includes change in soil moisture, reservoir storage, melting and consequent run-off from frozen ground in sub-Arctic areas, and changes in storage of water in vegetation.
- Contributions from glaciers and ice sheets that cover some 10 per cent of the Earth's land surface – as has already been identified, the widespread 20th-century retreat of valley mountain glaciers in non-polar regions is well documented and is estimated to contribute 0.2 to 0.4 mm yr^{-1} of sea-level change. Current estimates of the contribution (positive or negative) to sea-level change from the large polar ice sheets of Greenland and Antarctica are less well constrained.

Table 8.2   Estimated contributions to sea-level rise

Cause	20th-Century Contribution (mm yr^{-1})	Predicted 21st-Century Contribution (mm yr^{-1})
Thermal expansion	+0.3 – +0.7	+1.1 – +4.3
Melting of valley glaciers and ice-caps	+0.2 – +0.4	+0.1 – +2.3
Melting in Greenland	0.0 – +0.1	–0.2 – +0.9
Melting of Antarctica	–0.2 – 0.0	–1.7 – +0.2
Ice sheet adjustment since LGM	0.0 – +0.5	0.0 – +0.5
Changes in terrestrial storage of water	–1.1 – +0.4	–2.1 – +1.1
Sediment deposition	0.0 – +0.05	Small
Vertical land movements	Localised	Localised
IPCC total estimates	–0.1 – +1.9	+0.9 – +8.8

Source: IPCC, 2007

Unequal impacts of climate change

The previous section has highlighted the differences in projected impacts of the phenomenon that is global climate change, in which a 0.09–0.88 metre sea-level rise between 1990 and 2100 will have a much more dramatic impact on the population, geography and economy of a country in the Global South, such as Senegal or Bangladesh, compared to The Netherlands or Florida in the USA. An

Table 8.3   Impacts of the 2004 hurricane season in the Caribbean and south-east USA

Hurricane Event	Deaths	Insured Losses	Notes
Jeanne (USA, Haiti, Puerto Rico)	3034 (1700 in Haiti)	$4000 million	Haitian economy severely affected; social problems (e.g. looting)
Charley (USA, Cuba)	24 (none in Cuba)	$8000 million ($4.5 m worth of damage in Cuba)	In Cuba: 2 million people evacuated (17% of the population)
Frances (USA, Bahamas)	38	$5000 million	
Ivan (USA, Barbados, St Lucia, St Vincent, Grenada)	124	$11,000 million	Most costly insurance loss in 2004, including damage to oil rigs in the Gulf of Mexico

Source: SwissRe, 2006

indication of this can be seen in the string of hurricanes[1] affecting the Caribbean and south-east USA in the autumn of 2004. Table 8.3 shows the death toll and disruption caused by these events. Bearing in mind that the overall death toll in Florida for all four hurricanes was 113 and that no deaths occurred in Cuba (where two million people – 17 per cent of the population – were evacuated), you should be able to suggest both the nature of, and reasons for, these disparities.

Of course, these insurance losses were eclipsed in 2005 by Hurricane Katrina, which generated an estimated $45,000 million worth of insurance claims (SwissRe, 2006). This extreme event also drew attention to the fact that economic and social disparities do not just exist at the global level, but within countries also. The controversy over the disproportionate impact of Hurricane Katrina on poor, black communities in New Orleans is well known: with a pre–Katrina population of 485,000 in the city, around two-thirds were black. By the end of 2006, the city was half the size and only half the population was black, suggesting that African-Americans were most likely to have been affected, and not the first to return (see also pp. 227–229). In particular, the areas worst hit were low-lying neighbourhoods most at risk of flooding, and mostly inhabited by African-Americans. Chapter 9 on natural hazards develops these points. This draws attention to the fact that climate change impacts will affect individuals and communities, as well as states, in different ways.

There are many factors that contribute to such inequalities, the main one of which is poverty, but others include race, gender and age, as Chapters 2 and 3 have already illustrated with regard to environmental justice. Global and local income differentials (such as uneven terms of trade and uneven development) and their effects, such as poverty, need to be considered. Senegal and Bangladesh (and their neighbours in the Global South) have far fewer resources to mitigate climate change, or prevent it happening in the first place. For example, Haiti, which experienced the worst death toll of the 2004 Caribbean hurricane season, illustrated in Table 8.3 above, is one of the world's poorest countries, suffers badly from deforestation and has no disaster preparedness (Tearfund, undated). Such countries are poorly represented in organisations that are either controlled by the major contributors to global climate change (such as The World Bank and G8) or have the capacity to reduce its impact. Moreover, there are often irresistible pressures to develop in areas of vulnerability, which countries in the Global North are better equipped to avoid or to protect. A report commissioned by the new economics foundation has estimated that global warming will create 20 million environmental refugees a year, with 150 million people displaced by the impacts of global warming by 2050 (Conisbee and Simms, 2003), the overwhelming majority of whom will originate in the Global South. More recently, the Stern Review has presented research that calculates that climate change could generate as many as 150–200 million environmental refugees (that is, 2 per cent of the projected population) by the middle of the 21st century (Myers and Kent in Stern, 2006).

There are also differences in how effects of climate change may be experienced within countries: Box 8.6 explains the gender dimension of climate change which should get you thinking about how other groups (such as children, the infirm or disabled, as well as disadvantaged ethnic groups, as the New Orleans example above clearly illustrates) might experience this differently.

Box 8.6   Gender and climate change

In 1995, the United Nations 4th Conference on Women held in Beijing agreed that a 'gender perspective needed to be incorporated into all policies and programmes so that before decisions are taken, an analysis is made of the effects on women and men respectively'. This has been reiterated at a number of UN conferences subsequently and yet is a long way from being realised. As Box 8.7 below indicates, a recent Conference of the Parties to Kyoto has called for women to be nominated to sit on bodies established under the UNFCCC and the Kyoto Protocol. While women are very poorly represented on national governments and international bodies (mostly well below the 30 per cent required to create sufficient 'critical mass' to change policy in favour of women, Bhattar, 2001), they are frequently prominent among those who suffer the ill effects of climate change, or practices that exacerbate climate change. Consider the following:

- Women in the rural Global South are more likely than men to be in fuel poverty and, as deforestation gains pace, to have to walk increasing distances to collect fuel wood. This affects almost 40% of rural women in Latin America, almost 60% in Africa and nearly 80% in Asia (Bernstein, 2004).
- While there may be Western pressures on countries in the Global South to also reduce greenhouse gases, women in rural areas benefit significantly from having refrigerators and stoves, which respectively reduce disease, and the respiratory problems of cooking with wood, straw and husks (Bernstein, 2004).

(Continued)

- As climate change increases the likelihood of drought and endemic water shortage, this increases the time that women and girls spend on fetching water, which leads to girl children missing school (Denton, 2002).
- 80% of the global refugee community consists of women and children, and 75% of refugees from environmental and natural disasters are reported to be women and children. With increases in the refugee population as a result of environmental problems, this has a disproportionate impact on women and children. Women living in refugee camps are likely to spend an additional 20,000 extra woman hours a year over what they would have expected to spend at home, collecting water and fuel wood (Black, 1998; Stern, 2006).
- 90% of those who died in the 1991 cyclone in Bangladesh were women and children. This was largely as a result of their lack of education in disaster preparedness, and religious–cultural practices which isolated women from men (Fordham, 2003).
- Countries in Europe with greatest female political representation (such as Norway, Sweden and Germany) also tend to have the strongest environmental policies. The Chair of the Commission which investigated the relationship between environment and development (reported as Our Common Future in 1987) was the woman Prime Minister of Norway, Gro Harlan Brundtland. The European Commissioner for Environment in 1995–2004 was a Danish woman named Margaret Wallstrom (during a period in which the EU has taken a global lead on climate change negotiations).

Think about the possible reasons behind the practices illustrated by these statements, and identify some strategies that might mitigate the impacts of climate change without damaging women's lives disproportionately to men's.

The role of the IPCC and other institutions

While there now exists an intergovernmental framework for climate change (The United Nations Framework Convention on Climate Change – see Box 8.7), which has drawn up international legislation designed to reduce the production of greenhouse gases, the failure of the USA, Russia and Australia to ratify this meant that it was still not in force seven years after the Kyoto Protocol was agreed. Russia agreed to ratify the protocol late in 2004, which gave the protocol the number of states needed to bring it into force in February 2005. The path to getting to the situation, in which the necessary 55 per cent of UN member states have ratified the treaty, has been tortuous, and this can be followed from 1988 when the IPCC was set up to investigate climate change.

The drawn-out nature of coming to an agreement on modifying greenhouse gas emissions suggests that there are significant political ramifications. For example, the time period from the initial agreement to reduce greenhouse gases has seen four US national elections involving a change of government with somewhat different views on climate change from its predecessor. International politics may

Box 8.7   Climate change negotiations

Instruments/Dates	Achievements
1988 – Toronto Conference	Agreed 20% cut of greenhouse gases from 1988 levels by 2005. Intergovernmental Panel on Climate Change (IPCC) set up by UNEP and the World Meteorological Organization
1989 – 1st IPCC Assessment Report published	1st IPCC Assessment Report
1992 – United Nations Conference on Environment and Development, Rio de Janeiro	United Nations Framework Convention on Climate Change (UNFCCC) established and agreed to 'stabilise greenhouse gas concentrations in the atmosphere at a level that would prevent dangerous anthropogenic interference with the climate system. Also agreed to use the 'precautionary principle'. Ratified in 1993 and annual Conference of the Parties established
1994 – 1st Conference of the Parties (CoP), Berlin	Agreed that firm, and tougher, reduction targets were needed
1996 – 2nd IPCC Assessment Report published	Acknowledged for the first time that human influence on climate change seemed likely
1997 – 3rd CoP, Kyoto	Kyoto Protocol identified what needed to be done to reduce human-produced greenhouse gases: to reduce global emissions by 5.2% of 1992 levels by 2008/12; to show improvement by 2005; Annex I countries must achieve an 8% reduction; established potential for emissions trading; must be ratified by 55 parties accounting for at least 55% of total CO_2 emissions for 1990
2000 – 5th CoP, The Hague	Agreed that mechanisms for action were necessary, although no agreement reached between the USA and Europe on mechanisms
2001 – 6th CoP, Bonn	Agreed on the mechanisms for action: carbon sinks; international carbon trading; technology transfer to less developed countries; plans and targets to be submitted to CoPs
2002 – 7th CoP, Marrakesh	Agreed the terms of implementation. Also, for the first time, invited parties to give consideration to the active nomination of women to any body established under the UNFCCC or the Kyoto Protocol. Agreed to a consultation on technology transfer
2004 – Russia ratifies Kyoto Protocol	Kyoto Protocol entered into force in February 2005
2006 – 12th CoP, Nairobi (2nd Meeting of the Parties, MoP)	Agreed the 'Nairobi Framework', by which six United Nations agencies have launched an initiative to help developing countries – especially in Africa – participate in the Kyoto Protocol's Clean Development Mechanism. First discussions about the shape of an agreement to replace the Kyoto Protocol that expires in 2012

also be a more potent factor than science in determining what action takes place. For example, the UK Chief Government Scientist, Sir David King, warned the American Association for the Advancement of Science that climate change causes a bigger global threat than terrorism, for which he was reprimanded by former Prime Minister Blair (O'Riordan, 2004), although Blair himself argued this line in 2005, as the book's introduction points out. O'Riordan, in discussing this, suggests that Blair's concern was motivated by the UK's relationship with the USA in election year, confirming the highly politicised nature of climate change.

The Conferences of the Parties identified in Box 8.7 have been used to thrash out a number of issues, often with very limited success. For example, in The Hague in 2000, a compromise between the USA (trying to avoid or at least limit international legislation) and the European Union (which has been in the vanguard of legislating for greenhouse gas reductions) failed over the US insistence on the use of 'carbon sinks'. Nevertheless, by the Bonn conference a year later, the EU had conceded the use of carbon sinks, at which point the USA and Australia (another legislation resister) negotiated not only for new plantings to count as action on climate change, but new management practices of existing forests, and changed farming practices. Ultimately, as is now famously recorded, the USA pulled out of the Kyoto agreements on the basis that American interests were not best served by it, and arguing that countries outside 'Annex I' (the signatory countries in the West), such as China and India, should be required to reduce their emissions. That this coincided with the new presidency of George W. Bush underlined the influence that the oil and automotive industries wield in American politics.

That the USA had still not signed the Kyoto Protocol by 2007 represents a major barrier to international achievements to reduce climate change and moves to a low-carbon economy. Ironically, however, a number of US states, including the New England states, and California and Oregon, utilising their power as federated states, have independently signed up to the protocol and are introducing measures aimed at reducing their carbon footprint. Opinion may be turning in the USA: in 2006, a Democratic Party, more sensitive to the need to legislate to mitigate climate change, won control of the Senate and House of Representatives. The film 'An Inconvenient Truth', in which the ex–vice president Al Gore campaigns to raise the awareness of the problems of climate change, also appears to have had an impact on public and some industrial opinion.

This change is mirrored in the UK with the publication of the Stern Review in late 2006, which unequivocally states that 'the evidence gathered by the Review leads to the simple conclusion: the benefits of strong, early action considerably outweigh the costs' (Stern, 2006: ii). The year 2006 appears to have been pivotal with regard to the acknowledgement of climate change as an overwhelmingly critical issue, and it is interesting to contemplate why this may be so. The publication of the Stern Review and the opening of 'An Inconvenient Truth' were both significantly heralded in the media. Politicians have been increasingly eager to align themselves with the issue: witness David Cameron, as leader of the Conservative Party, being filmed examining the retreat of Arctic glaciers in Norway ahead of the 2006 political party conferences, all of which held a 'Climate Clinic' fringe event, put together by an alliance of environmental groups in the UK (*The Independent*, 2006). 'Global warming' received an accolade of style in May 2006 when it hit the covers of *Vanity Fair* magazine as referred to in Chapter 3.

It is no longer surprising to hear business leaders express concern about climate change: from Richard Branson deciding to invest £3 billion in biofuel development for the airline industry (see Chapter 1) to the Confederation of British Industry launching a task force in January 2007 to discuss ways in which companies such as BT, Tesco and BA can tackle climate change, including methods such as carbon taxes and offsetting. According to BBC business editor Robert Peston, 'some of Britain's biggest companies are now admitting climate change is real, dangerous and partly their fault' (Peston, 2007). Despite this, however, national and international policy is still heavily influenced by the agendas of powerful commercial interests, which the next section addresses.

Box 8.8   Policy terms

Carbon sinks: these are biological resources that can absorb carbon, such as forests, grasslands and oceans. The burning of forests, therefore, not only produces CO_2 in its own right, but also reduces the capacity of nature to absorb CO_2.

Carbon taxes: taxes imposed on carbon producing activities. For example, in the UK, car tax is now scaled to the amount of CO_2 a vehicle emits. The owner of a low CO_2 emission vehicle will pay less than the owner of a higher emitting vehicle.

Carbon trading: this enables producers of CO_2 to trade their quotas; thus, one producer not wishing to reduce emissions may buy permits from another who is introducing cleaner technology to reduce its emissions. Trading can take place between industries (brokered through a 'carbon exchange' such as that in Chicago), between governments, or between individuals (see, for example, the RSA's campaign to introduce personal carbon trading).

Carbon offsetting: Defra (2007) describes this as involving a calculation of individual emissions and then purchasing 'credits' from emission reduction projects, which claim to prevent or remove an equivalent amount of carbon dioxide elsewhere, such as by investing in energy efficiency projects or afforestation programmes. These schemes are quite contentious with some environmental organisations, such as Friends of the Earth, suggesting that their value is limited, as they allow people to continue with their resource-intensive lifestyles, rather than change their behaviour to be more environmentally sustainable.

Clean development mechanism: under the Kyoto Protocol, the main formal channel for supporting low carbon investment in developing countries. Governments and the private sector can invest in projects that reduce emissions in fast-growing emerging economies.

Precautionary principle: a requirement on the polluter to take action to reduce likely pollution. This approach replaces an earlier approach that required those suffering pollution to prove its ill effects.

(Continued)

Reinsurance: insurance companies are finding it increasingly difficult to survive the increasing number of extreme disasters that substantially increase the number of insurance claims. Reinsurance effectively insures the insurers, so that insurance companies can spread their risk. Consequently, when a big insurance claim hits an insurance company, it is able to meet its obligations because it has spread its own risk around a number of reinsurance companies, such as SwissRe and MunichRe.

Technology transfer: this involves the sharing of technology developed in Annex I countries, which will enable less developed countries to introduce clean technology. Agreed under the Marrakesh Accords (see Box 8.7), the UNFCCC has agreed a framework to assess technology needs, establish a technology information system, create environments for technology transfer, provide the capacity building necessary to enable the transfer and provide funding to implement the framework.

Sceptics or visionaries?

Who pays for the research? Can researchers be bought? Are there controls on what researchers are allowed to research into and publish? Climate change science is a social construction that cannot be disentangled from political biases, interpretations and expectations of funders and regulators (Jasanoff, 1990; Wynne, 1994). Both scientific research and monitoring, let alone the integrated assessment process, are triggered by political values and ideological conflict (Jäger and O'Riordan, 1996).

It will come as no surprise, therefore, that business interests collude to finance scientific interpretations that are contrary to established IPCC viewpoints. The American Petroleum Institute, for example, has cooperated with the coal industry to review and critique global circulation models (O'Riordan and Jäger, 1996), while the Global Climate Coalition, set up in 1989 primarily by the oil industry, was formed to provide an alternative view on climate change which would favour the fossil fuel industry, and this has lobbied governments to develop policy sympathetic to its concerns. Another example of the power of industrialists is found in Exxon–Mobil which succeeded in having the chair of the Intergovernmental Panel on Climate Change (IPCC) replaced by someone they found more amenable to the oil and gas industry (O'Riordan, 2004).

In January 2007, Chrysler's chief engineer, Van Jolissant, described climate change as 'way way in the future, with a high degree of uncertainty'. A spokesman from DaimlerChrysler was then reported in the media as stating that 'while the science of climate change remained "uncertain", the company supported concurrent advances in climate science to ensure a fuller understanding of the controversies surrounding this issue and to avoid inappropriate responses by government or private sector' (BBC, 2007).

This tension between the different viewpoints and their supporters was thrown into relief by the broadcasting of 'The Great Global Warming Swindle' in the UK and Australia in 2007, in which a group of climate change sceptics claimed that scientists were at best wrong, or at worst lying, about the extent of anthropogenic climate change. The film provoked heated debate, and was much criticised by eminent panels of scientists in both countries, on the basis of poor evidence underpinning the programme, scientific errors and editorial bias (Climate of Denial, 2007; Jones et al., 2007). The only climate change scientist recruited to the programme claimed that he had been misrepresented, and the programme's director has a record of producing misleading environmental broadcasts that have had to be retracted later (Monbiot, 2007).

Another exercise which has attracted controversy has been the 'Copenhagen Consensus' organised by Bjorn Lomborg (author of *The Skeptical Environmentalist* referred to at the beginning of the chapter), in which he assembled 'eight of the world's most distinguished economists' (Copenhagen Consensus, undated) to rank a range of global problems including climate change, by their responsiveness to investment. Climate change was ranked last. This exercise begs a number of questions which could include: the basis on which the problems were ranked and which eight of the 'world's most distinguished economists' were selected. The criteria for ranking was a cost–benefit rationale, which Chapter 4 has already identified as being problematic for a range of reasons, and when the backgrounds of the economists are considered, it is notable that there is a strong free-trade/libertarian bias and no ecological economists. For example, Vernon Smith is a fellow of the right-wing American 'Cato Institute' which is committed to 'libertarianism, individual liberty, free markets and peace' (Cato Institute, 2007); Jagdish Bhagwati is an advocate of free trade; Thomas Schelling is on record as believing that climate change is an exaggerated threat for the USA (although of serious consequence for the developing world); and Robert Fogel is well known for a controversial analysis of antebellum slavery in the USA, which claimed it was 'a lucrative, robust and rational economic system . . . 35 per cent more efficient than Northern family farms' (Gibson, 2007) – a conclusion which arguably points up the limitations of cost-benefit analysis. The Copenhagen Consensus, then, is a good example of the need to examine the provenance of ideas and claims concerning climate change, and a caution against accepting such claims at face value.

Dealing with the problem

Ultimately, there are two key ways of dealing with the problem of climate change – mitigating against future change or adapting to the resultant change.

Mitigation

Mitigation initiatives approach the problem of climate change by reducing emissions of greenhouse gases, such that likely future rises in greenhouse gases' atmospheric concentrations will be reduced, or reversed. As stated in Article 3.3 of the Framework Convention on Climate Change: 'The parties should take precautionary measures to anticipate, prevent or minimise the causes of climate change and mitigate its adverse effects' (UNFCCC, 1994). These analyses depict mitigation efforts as a type of insurance against potentially serious future consequences. A weak version of this precautionary principle is the idea that, in the presence of uncertainty, it may be prudent to engage in policies that provide insurance against some of the potential damages from climate change. In its stronger form, the precautionary principle stipulates that nations should pursue whatever policies are necessary to minimise the damages under the worst possible scenario. This stronger form assumes extreme risk–aversion, since it focuses exclusively on the worst possible outcomes (Lyon, 2003). Such principles can take the form of voluntary or regulated actions.

Lyon (2003) classifies voluntary programmes into three broad categories: unilateral initiatives, negotiated agreements and public voluntary agreements (PVAs). Unilateral initiatives by industry are sometimes referred to as self–regulation and are typically seen as attempts to ward off regulatory threats. They may also produce a range of ancillary benefits, however, including wooing environmentally sensitive consumers and investors, and influencing future regulatory programmes. Self-regulation can avoid the costly process of passing legislation and implementing regulations, and give industry the flexibility to meet environmental goals in a cost-efficient manner. Negotiated agreements are also a means of averting a regulatory threat, but the government, by participating in the negotiation process, can make a clear policy statement and can push industry to go beyond what it would have done on its own. Finally, PVAs are programmes involving government provision of technical assistance in meeting environmental goals, government-sponsored publicity for firms with outstanding environmental records, and information sharing between participating firms. Often, this will result in government agencies including environmental statements in tenders, ensuring environmental goals are met through the economic process.

A plethora of local government initiatives and legislation is attempting to change consumer attitude, such that emissions are reduced at an individual level as well as at the corporate level. For example, mitigation options in the automotive sector include: more fuel-efficient vehicles; the development of alternative fuel sources such as biofuels, and electric and hybrid vehicles with more powerful and reliable batteries; taxation or road tolling for high carbon-emitting cars; lower road taxes for energy-efficient cars; investment in rail and inland waterway shipping to produce a shift away from road use; and encouragement to move from low-occupancy to high-occupancy passenger transportation through the provision of higher-standard, mass transport options.

Adaptation

Many early impacts of climate change can be effectively addressed through adaptation. The array of potential adaptive responses available is very large, ranging from purely technological (e.g. sea defences), through behavioural (e.g. altered food and recreational choices) and managerial (e.g. altered farm practices), to policy (e.g. planning regulations). Many social and economic systems, including agriculture, forestry, settlements, industry, transportation, human health, and water resource management, have evolved to accommodate some deviations from 'normal' conditions. This adaptation rarely includes the extreme events predicted with climate change. Adaptation is also often reactive (undertaken after impacts are apparent) rather than anticipatory (undertaken before impacts are apparent).

Early planning for the impacts of climate change is likely to bring considerable advantages. Many decisions made today will have consequences for decades. It is cheaper, for example, to design new housing or infrastructure to cope with a future climate than to retrofit later. A systematic approach to adaptation planning, such as risk management, will help identify information needs (IPCC, 2007).

The options for successful adaptation diminish and the associated costs increase with increasing climate change. At present, we do not have a clear picture of the limits to adaptation, or the cost, partly because effective adaptation measures are highly dependent on specific, geographical and climate risk factors, as well as institutional, political and financial constraints. As for mitigation, a wide array of adaptation options is available, but more extensive adaptation than is currently occurring is required to reduce vulnerability to future climate change, and adaptation alone is not expected to cope with all the projected effects of climate change.

It is noteworthy that the Stern Review referred to earlier has concluded that the human and economic costs of adaptation are of such consequence that mitigation '... – taking strong action to reduce emissions – must be viewed as an investment, a cost incurred now and in the coming few decades to avoid the risks of very severe consequences in the future' (Stern, 2006: i).

Summary

This chapter has illustrated some of the uncertainties behind the science of climate change: the inevitable incompleteness of scientific knowledge; the complexity of the very many components in climate change; and the unpredictability of these components, and even more of the synthesis of these components. It has also explored the nature of political and business interests that seek to influence any attempt to regulate factors contributing to climate change, in order to protect their own interest.

As this chapter was being written, there have been a number of claims that climate change is becoming increasingly out of control (McGuire 2004), and concerns expressed that it is of a completely different magnitude compared to other problems facing humanity (Monbiot, 2007). On the other hand, as the chapter has already illustrated, the Copenhagen Consensus has dismissed climate change as 'the least important of the world's immediate problems' and as such not worthy of investment.

What is clear and unarguable is that whatever effects are being felt, and their relative importance or unimportance, these effects are unevenly spread both at the macro (global) and at the micro (as in the community) level. Moreover, there is a clear inverse relationship between those contributing the most to anthropogenic climate change and those experiencing most of the negative effects. What should be done about this is the debate that is laboriously being worked out at the national and global level, and it is a position that a student of environmental issues in general, and climate change in particular, should consider.

At the time of this book's publication, there appears to be an increasing agreement that climate change is, indeed, a major world problem, demanding attention and action. The reasons for this change in perspective are not entirely clear, but are likely to include the nature of the data itself; the high profile of key authorities and their champions, including media stars; the work of high-profile NGOs on climate change (such as Friends of the Earth and Greenpeace); the increasing availability of alternatives to fossil fuels (such as wind and solar power, biofuels and hydrogen), which become more economically feasible and offer profitable opportunities for business; and the high profile of extreme events which, while their link with climate change might not be scientifically conclusive, are persuasively linked to climate change in people's minds.

Note

1. While these severe weather events – creating the highest insured damage of any hurricane season to that date – have been linked to global warming in the popular media, there is debate over the extent to which they are linked, as this chapter has already indicated. However, it is interesting to note that SwissRe, the reinsurance company from which the data is drawn, is now suggesting a link between climate change and severe weather events; likewise, the Stern Review proposes that 'extreme weather events are likely to occur with greater frequency and intensity in the future, particularly at higher temperatures' (Stern, 2006: 132)

References

BBC (2007) 'Chrysler question climate change', http://www.news.bbc.co.uk/1/hi/business/6247371.stm (accessed 10 January 2007).

Bernstein, J. (2004) 'Promoting gender equality, providing energy solutions, preventing climate change', seminar report to Conference of the Parties 9, Milan, 9 December 2003. Stockholm: Swedish Ministry for the Environment.

Bhattar, G. (2001) 'Of geese and ganders: mainstreaming gender in the context of sustainable human development', *Journal of Gender Studies*, 10(1): 17–32.

Black, R. (1998) *Refugees, Environment and Development*. London: Longman.

Cato Institute (2007) http://www.cato.org/ (accessed 16 August 2007).

Climate of Denial (2007) http://www.climateofdenial.net/ (accessed 14 August 2007).

Conisbee, M. and Simms, A. (2003) *Environmental Refugees: The Case for Recognition*. London: new economics foundation.

Copenhagen Consensus (undated) 'Copenhagen Consensus: the results', http://www.copenhagenconsensus.com/Admin/Public/Download.aspx?file=Files/Filer/CC/Press/UK/copenhagen_consensus_result_FINAL.pdf (accessed 16 August 2007).

Defra (2007) http://www.defra.gov.uk/environment/climatechange/uk/carbonoffset/ (accessed 14 August 2007).

Denton, F. (2002) 'Climate change vulnerability, impacts and adaptation: why does gender matter?', in R. Masika (ed.), *Gender, Development and Climate Change*. Oxford: Oxfam.

Douglas, B.C. (1991) 'Global sea–level rise', *Journal of Geophysical Research*, 96(C4): 6981–92.

Fordham, M. (2003) 'Gender, disaster and development: the necessity for integration', in M. Pelling (ed.), *Natural Disasters and the Developing World*. London: Routledge.

Gibson, L. (2007) 'The human equation', *University of Chicago* magazine, 99: 5. http://magazine.uchicago.edu/0726/features/human.shtml (accessed 16 August 2007).

Holligan, P.M. and deBoois, H. (eds) (1993) Land–ocean interactions in the coastal zone (LOICZ), science plan. Stockholm: International Geosphere Biosphere Programme (IGBP), International Council of Scientific Unions.

Houghton, J. (2004) *Global Warming: The Complete Briefing*. Cambridge: Cambridge University Press.

The Independent (2006) Climate Clinic supplement, 2 October.

IPCC (Intergovernmental Panel on Climate Change) (1996) *The IPCC Second Assessment Report: Climate Change 2001*. Cambridge: Cambridge University Press.

IPCC (Intergovernmental Panel on Climate Change) (2001) *The IPCC Third Assessment Report: Climate Change 2001*. Cambridge: Cambridge University Press.

IPCC (Intergovernmental Panel on Climate Change) (2007) *The IPCC Fourth Assessment Report: Climate Change, 2007* at http://www.ipcc.ch/ (accessed 16 August 2007).

Jäger, J. and O'Riordan, T. (1996) 'The history of climate change science and politics', in T. O'Riordan and J. Jäger (eds), *Politics of Climate Change: A European Perspective*. London: Routledge.

Jasanoff, S. (1990) *The Fifth Branch: Scientific Advisors as Policy Makers*. Cambridge, MA: Harvard University Press.

Jones, D., Watkins, A., Braganza, K. and Coughlan, M. (2007) '"The Great Global Warming Swindle": a critique', *Bull. Aust. Meteor. Ocean. Soc.*, 20(3): 63–72.

Kiehl, J.T. and Trenberth, K.E. (1997) 'Earth's annual global mean energy budget', *Bulletin of the Meteorological Society*, 78(2): 197–208.

Lockwood, M. and Fröhlich, C. (2007) 'Recent oppositely directed trends in solar climate forcings and the global mean surface air temperature', *Proc. R. Soc. A.*, doi:10.1098/rspa.2007.1880.

Lomborg, B. (2001) *The Skeptical Environmentalist*. Cambridge: Cambridge University Press.

Lyon, T.P. (2003) 'Voluntary versus mandatory approaches to climate change mitigation', issue brief, Resources for the Future, Washington DC.

MacCracken, M. and 18 others (1991) 'Working Group 2: a critical appraisal of model simulations', in M.E. Schlesinger (ed.), *Greenhouse-Gas-Induced Climatic Change: A Critical Appraisal of Simulations and Observations*. New York: Elsevier.

McGuire, W. (2004) 'Climate Change 2004' *Technical Paper 02*. London: Benfield Hazard Research Centre, University College.

Monbiot, G. (2007) 'Don't let truth stand in the way of a red-hot debunking of climate change', *Guardian Unlimited*, 13 March, http://www.guardian.co.uk/comment/story/0,,2032361,00.html (accessed 14 August 2007).

Nicholls, R.J. and Leatherman, S.P. (1995) 'The implications of accelerated sea-level rise for developing countries: a discussion', *J. Coast Res.*, 14: 303–23.

O'Riordan, T. (2004) 'Environmental science, sustainability and politics', *Transactions of the Institute of British Geographers*, 29(2): 234–47.

O'Riordan, T. and Jäger, J. (1996) *Politics of Climate Change: A European Perspective*. London: Routledge.

Peston, R. (2007), news.bbc.co.uk/1/hi/business/6250763.stm

Radford, T. (2003) 'Global warming endangers Amazon', *The Guardian*, 17 February.

Shackley, S., Young, P., Parkinson, S. and Wynne, B. (1998) 'Uncertainty, complexity and concepts of good science in climate change modelling: are GCMs the best tools?', *Climatic Change*, 38: 159–205.

Shepherd, J.M. and Knutson, T. (2007) 'The current debate on the link between global warming and hurricanes', *Geog. Compass*, 1(1): 1–24.

Stanhill, G. (2001) 'The growth of climate change science: a scientometric study', *Climatic Change*, 48: 515–24.

Stern, N. (2006) *Stern Review: The Economics of Climate Change.* London: HMG.

SwissRe (2006) http://www.swissre.com/ (accessed 1 August 2006).

Tearfund (undated) *Learn the Lessons.* Teddington: Tearfund. http://www.tearfund.org/webdocs/Website/News/Disasters%20Media%20Report%20-%20SMALLER%20VERSION.pdf (accessed 16 August 2007).

UNFCCC (1994) 'The United Nations Framework Convention on Climate Change', http://unfccc.int/essential_background/convention/background/items/1355.php

UNFCCC (2003) *Caring for Climate: A Guide to the Climate Change Convention and the Kyoto Protocol.* Bonn: UNFCCC.

United Nations (2002) 'WSSD Plan of Implementation'. New York: United Nations.

Vaughan, D.G. and Spouge, J.R. (2002) 'Risk estimation of collapse of the West Antarctic Ice Sheet', *Climatic Change*, 52: 65–91.

Wynne, B. (1994) 'Scientific knowledge and the global environment', in M. Redclift and T. Benton (eds), *Social Theory and the Global Environment.* London: Routledge. pp. 168–89.

9 Natural Hazards

Iain Stewart and Katherine Donovan

Learning outcomes

Knowledge and understanding of:

- ○ what is meant by the term 'natural hazard'
- ○ the geographical distribution of hazards
- ○ the distinction between 'natural' hazards and human created 'disasters'
- ○ the terms 'vulnerability', 'risk' and 'resilience'

Critical awareness and evaluation:

- ○ you will have learnt that the variations in the effects of disasters are directly related to global and local differentials in social cohesion, economic development, wealth and governance, and be able to evaluate these processes
- ○ you will be able to evaluate similar variations in post-disaster recovery
- ○ you will be able to reach critical judgements in relation to disasters and hazards and develop a personal standpoint based on the evidence presented in this and other works on the subject

---

Natural hazards – a growing threat

'In a time of extraordinary human effort to live harmoniously in the natural world, the global death toll from extreme events of nature is increasing. Loss in property from natural hazards is rising in most regions of the earth, and loss of life is continuing or increasing among many of the poor nations of this world.'

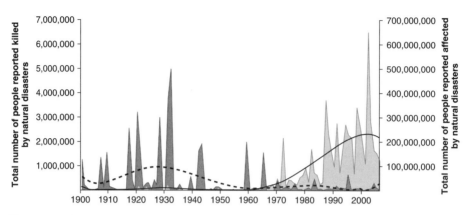

Figure 9.1   Human losses due to natural disasters, 1900–2006. Dark shading indicates fatalities reported due to natural disasters (scale on left) and light shading indicates total number of people affected (scale on right). Dashed line shows the smoothed trend for fatality numbers and solid line shows smoothed trend for number affected

Source: EM-DAT – the OFDA/CRED International Disaster database, http://www.em-dat.net, Université Catholique de Louvain, Brussels, Belgium.

These words were written almost 30 years ago, the opening lines of Burton, Kates and White's now-classic text *The Environment as Hazard* (Burton et al., 1978: 1), but they are as equally apt today. Despite three decades of technological advance, and including a well-intentioned decade (1990–2000) devoted to the natural disaster reduction (Press and Hamilton, 1999), events such as earthquakes, volcanic eruptions, tropical cyclones, floods and droughts continue to inflict significant fatalities and soaring economic and social costs. In the last two decades alone, more than 1.5 million people have been killed by natural disasters (EM-DAT, 2006), and though levels of fatalities appear to be on the decline, the ratio of persons affected per persons killed has risen steadily through the 20th century (Figure 9.1) (EM-DAT, 2006).

Of the 9000 disasters reported since 1900 (EM-DAT database), 80 per cent have occurred in the last 30 years. During that latter period, there has been a four-fold increase in the annual number of reported natural disasters, from fewer than 100 in the mid-1970s to a little more than 400 in 2003 (Guha-Sapir et al., 2004). No doubt this partly reflects a significant increase in the reporting of small disasters by voluntary organisations and humanitarian agencies and recorded in databases that have appeared in recent decades, such as those maintained by the Center for Research on the Epidemiology of Disasters (CRED), the Munich Reinsurance company, and the Latin American network La Red (Dilley, 2006). But behind this escalating disaster toll lies a complex set of societal changes: the increase in the world's population, their concentration in large conurbations, social and economic consequences of development in highly exposed regions, and the high vulnerability of modern societies and technologies (Smolka, 2006). In simple terms, more people and more assets are being placed in harm's way.

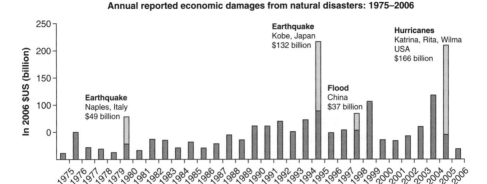

Figure 9.2   Annual reported economic damages from natural disasters, 1975–2006. Selected disasters with significant economic impact shown in lighter shading. Epidemics and insect infestations not included

Source: EM-DAT – the OFDA/CRED International Disaster database, http://www.em-dat.net, Université Catholique de Louvain, Brussels, Belgium

The geography of hazard

Hazards are not 'bolts from the blue' – by and large, scientists can forecast where they will occur because they have recurred repeatedly in the same places. Droughts have been occurring in the Sahelian region of Africa for millennia, monsoonal storm surges annually inundate the delta flats of Bangladesh, and hurricanes strike with reasonable predictability in the Atlantic and Gulf coasts of the USA. These natural agents are not novelties where they occur, but periodic regularities – systemic elements of specific environmental circumstances. Consequently, there is a geography to the risks that such agents bring. Mortality associated with geophysical hazards (earthquakes, volcanic eruptions, tsunamis) is highest along geological plate boundaries around the Pacific Rim and across southern Asia, storm and flood hazards have greatest lethality across the sub-tropical zones, and drought-related fatalities are concentrated in the semi-arid regions of Africa (Dilley et al., 2005).

But there is a deeper geographical bias to natural hazards. About 97 per cent of hazard-related deaths, and 99 per cent of people affected by natural emergencies, occur in developing countries (Twigg, 1998). This is also where the greatest financial burden of natural hazards lies (Dilley et al., 2005). Average income losses from recent disasters in some developing countries (e.g. Sri Lanka, Bangladesh, Nicaragua) are ten to twenty times greater than from disasters in more developed nations (Haas et al., 1977). Floods and droughts that claim about one-tenth of 1 per cent of the Gross Domestic Product (GDP) of industrialised countries, cost up to twenty times more (up to 2 per cent of GDP) in less developed nations (Alexander, 1993). Thus, while the financial costs of natural disasters in industrialised countries have, in the long term, comparatively little effect on national economies, many Third World nations will be hard pressed to develop economically due to recurrent hazard losses. For many countries, probable economic losses over the next century exceed their current financial resources (e.g. Cardona, 2005). Lacking the wealth, infrastructure and institutional capacity enjoyed in other parts of the world, they cannot afford the same levels of protection as more developed countries, which have invested substantially in a wide range of preparedness and mitigation measures.

And yet it is precisely because of that investment that the world's wealthiest nations now hold the greatest economic assets and, consequently, suffer the costliest disasters in history (Dilley et al., 2005). In Europe, North America and Japan, improvements in scientific forecasting, safer buildings, regulations on the use of land, and extensive emergency management systems have greatly reduced the number of disaster-related fatalities over recent decades, but economic losses from hazards have witnessed a several-fold increase. In the USA, for example, the economic losses from hazards during the mid-1990s were costed at US$54 billion per year – or a staggering $1 billion per week (van der Vink et al., 1998). What is telling is that it is those states most affected by the costs of hurricanes (Florida, Maryland, North Carolina, and Texas) and earthquakes (California and Washington) that show the largest increases in population and revenue (van der Vink et al., 1998). In other words, more people are moving to the USA's hazard-prone areas, and they tend to be the more affluent segment of American society. The message is clear: 'We are becoming more vulnerable to natural disasters because of the trends in our society rather than those of nature' (van der Vink et al., 1998: 537).

Counting the cost of natural disasters may be a tricky business (e.g. Alexander, 1993; Downton and Pielke, 2005), but by most measures, hazard impacts are on the rise. As the global economy grows, so the number and cost of natural disasters rises. This association suggests that economic growth has not been properly directed at mitigating disaster risks (Pérez-Maqueo et al., 2007). That is largely because the ensuing disasters are still viewed by many policymakers as 'exceptional natural events' (and, in some contexts, even as 'Acts of God') that temporarily disrupt 'normal' human development and therefore require extraordinary humanitarian action. In fact, as we will discover in the following sections, hazards are remarkably unexceptional events.

What do we mean by natural hazards?

Natural hazards arise out of the unceasing transformation of our physical environment by energy supplied from two unrelated sources of power. One emanates from heat loss from within the Earth, an internal convective engine that fuels the unceasing movement of tectonic plates – generating earthquakes and volcanic eruptions – and drives mountain building, from which unstable debris is shed via landslides. The other source of power that drives change emanates from the Sun, pertains to climate, and is manifest in the convective interactions between ocean and atmosphere that produce extreme weather events – hurricanes and cyclones, storm surges, heatwaves, floods and the like.

Because of their separate origins, natural hazards arising from tectonic or ocean–atmosphere interactions ought to occur independently of one another in time. In other words, the frequency with which earthquakes strike is essentially unaffected by the frequency with which, say, hurricanes or heatwaves occur. This temporal intermittence of natural crises generally allows most afflicted populations to respond and to recover during periods of relative quiescence. Occasionally, however, rather like intermeshing biorhythms, discrete geophysical hazards conspire to strike concurrently or in close succession. The major (M 7.7) earthquake that killed tens of thousands of people in the north-west Indian state of Gujarat in January 2001, for example, came at the end of a decade when the province had been hit by two damaging cyclones, a malaria epidemic, flooding and a prolonged drought. When two or more calamitous events affect a region at or around the same time, the potency of each disruptive episode magnifies the compound stress. These 'convergent catastrophes' (Moseley, 1999) can be visualised as the punctuated reduction in the adaptive flexibility of a human ecosystem as a consequence of repeated disaster blows (Dyer, 2002).

Convergent catastrophes are made more likely by the fact that there are complex interlinkages between geophysical systems. For instance, in October 2004, a series of powerful earthquakes triggered thousands of landslides in the Honshu district of northern Japan, but the effect of the strong seismic shaking was made far worse by antecedent rainfall conditions induced by a typhoon (Wang et al., 2006). Such contagion exacerbates hazard impacts, as does the frequency with which natural crises strike. As we shall explore later, this may be counteracted somewhat by the tendency for more regular hazards to inculcate cultural adaptation. There is also an important cultural distinction between rapid-onset disruptions – sudden high-intensity events like earthquakes and storms – and slow-onset disturbances like droughts and soil erosion, which unfold, unperceived, in increments over considerable amounts of time, only to be recognised well after their initial manifestation. In general, there has been a failure to recognise the cumulative toll of small impact events, which collectively probably cause more death and incapacitation than large-scale events (Hewitt, 1983). Not only can the cumulative effects of protracted hazards match or exceed those of short-term emergencies, but their long duration increases the probability of overlap with swift disasters to produce convergent catastrophes (Moseley, 2002).

Unnatural hazards

Earthquakes, hurricanes, floods, volcanic eruptions and other destructive 'geophysical' agents are, on the face of it, largely outside the control of human activity. In this regard, these 'natural' crises are generally considered to be distinct from anthropogenic emergencies (e.g. careless disposal of waste products, the release of dangerous substances into the atmosphere, or 'accidents' in nuclear-energy plants or biotechnical laboratories) that can also claim many lives. However, the distinction between 'natural' and 'technological' hazards is becoming distinctly blurred because humans don't simply occupy the physical environment, they modify it. Floods, for example, can be produced by the structural failure of dams or the inability of artificial drainage systems to withstand 'normal' deluges. For instance, a huge wave of mud that engulfed low-income neighbourhoods of Algiers, capital

Figure 9.3 The dynamic landscape of the Andes mountains – regularly afflicted by earthquakes, volcanic eruptions, landslides, floods and droughts, many of them resulting in convergent catastrophes that threaten major societal collapse (photo: the authors)

of Algeria, in November 2001, killing over 500 people, was the result of an urban drainage system that could no longer cope with flash floods; a few years earlier, the government had sealed off an extensive storm drainage system for fear of terrorists using it as a subterranean hide-out (Wisner, 2003). Cities themselves are becoming as dangerous as the natural environments they replace. Thus, a slide of solid waste from the Payatas rubbish dump in Manila (Philippines) in July 2000, which killed 300 people in the contiguous squatter settlement, is a true urban catastrophe for the 21st century (Pelling, 2002).

Many catastrophes traditionally attributed to natural causes are now more likely to be viewed as being generated, at least in part, by human practices, especially those connected to current conditions of environmental degradation. Much of the flood threat in Bangladesh, for example, has been ascribed to deforestation and hill cultivation in Nepal, releasing massive volumes of eroded sediment that clog the downstream sections of the Ganges–Bramaputra–Meghna drainage courses, making them more flood-prone (Eckholm, 1975; Myers, 1986). Recently, however, this Himalayan deforestation–hazard linkage has been rejected by Gardner (2003), who identifies population growth and economic development as the key culprits in accentuating hazard vulnerability. Here too, a tight coupling between political and natural hazards is evident. During the 2000 and 2004 floods which affected the India–Bangladesh border, Indian border security forces breached river embankments to allow the water to spill out, thereby ameliorating its downstream impacts

Figure 9.4 Geophysical hazards like volcanic lava flows, here destroying a house on Mount Etna, Sicily, are largely outside the control of human activity, yet many hazard agents are far from natural (photo: the authors)

in West Bengal (India), but exacerbating destruction of life, crops and property in Bangladesh (Ali, 2007).

The 'human' versus 'natural' dichotomy is further confused by the growing likelihood that human-induced climate changes are modifying the incidence and severity of some physical hazards, especially tropical storms (e.g. Mitchell et al., 2006; Munich Re, 2006; Trenberth, 2005), but also other destructive agents (e.g. Fengqing et al., 2005; Nelson et al., 2002). While the evidence for climatically driven changes in disaster frequency is inconclusive, the IPCC (2001) stated that '. . . it is likely that there has been a widespread increase in heavy and extreme precipitation events in regions where total precipitation has increased, e.g. in the mid- and high latitudes of the northern hemisphere'. Climate projections predict more extreme weather variability (e.g. Greenough et al., 2001), and it seems certain that future environmental change, through soil erosion, groundwater contamination, land subsidence and sea-level rise, will significantly increase the exposure of people to natural hazards (Adger and Brooks, 2003).

Of course, a human dimension to natural hazards has long been well accepted. After all, it is only when the natural processes threaten loss of life, damage to property, or disruption of individual or community routines or organisational structure, that we see the environment as hazardous at all. But, increasingly, many would argue that while physical exposure to floods, hurricanes, earthquake or volcanic eruptions is an important part of what makes individuals and communities hazard-prone, a more important part is people's lack of capacity to avoid, cope with, and recover from adverse events (e.g. Blaikie et al., 1994). In the following sections, we explore the notion that while hazards happen, disasters are caused – incurred by a lack of preparedness within communities, and by the inabilities of political authorities to organise and provide resources to guard against or withstand hazard effects. In this sense, there is little that is 'natural' about hazards.

Box 9.1   Hazard definitions

- **Disaster**: a serious disruption of the functioning of a community or a society, causing widespread human, material, economic or environmental losses which exceed the ability of the affected community or society to cope using its own resources (ISDR, 2004; Smith, 2001).
- **Hazard**: a potential threat to humans and their welfare (Smith, 2001).
- **Natural hazard**: a natural process or phenomenon occurring in the biosphere that may constitute a damaging event (ISDR, 2004).
- **Vulnerability**: the characteristics of a person or group in terms of their capacity to anticipate, cope with, resist and recover from the impact of a natural hazard (Wisner et al., 2004). The ISDR (2004) definition clarifies the conditions which determine these characteristics as physical, social, economic and environmental factors which increase the susceptibility of a community to the impact of hazards.
- **Risk**: the probability of loss resulting from the interaction of a hazard (earthquake, flood, volcanic eruption, etc.) and a vulnerability (community, infrastructure and economic activity):

$$Risk = Hazard \times Vulnerability$$

Risk is described in qualitative terms such as high risk and low risk.
- **Resilience**: is a measure of a community's ability to absorb and recover from the occurrence of a hazardous event.

Vulnerability: turning hazards into disasters

Hazards emerge from nature, but the disasters they inflict do not. 'Society, rather than nature, decides who is more likely to be exposed to dangerous geophysical agents' (Hewitt, 1997: 141). Around the world, circumstances conspire to make people live and work in hazardous locations: flood-prone marginal lands, precarious mountain slopes, fertile volcanic flanks, or earthquake-prone shores. Each of these environments exposes people to a bewildering amalgam of lethal threats. But even in these unsafe environments, many natural phenomena would not constitute 'hazards' if it wasn't for the inherent conditions of underdevelopment in which people have been forced to live (e.g. Pelling, 2003). In such circumstances, vulnerabilities emerges from the 'normal' order of things – they simply compound the struggles that are part of people's daily lives (Blaikie et al., 1994; Hewitt, 1983, 1997; Wisner et al., 2004), as this example demonstrates:

> On the eve of Bangladesh's massive floods in August 1988, this relatively powerless group [landless squatters] was living in an economically marginal situation but close to the city, on low-lying land prone to flooding. Their economic and political marginality meant they had few assets in reserve. It also meant that their children were unusually malnourished and chronically ill. This channelled the dynamic pressure arising out of landlessness and economic marginalisation into

Figure 9.5 In many parts of the world, vulnerability emerges from the 'normal' order of things – hazards simply compound the struggles that are part of people's daily lives (photo: the authors)

a particular form of vulnerability: lack of resistance to diarrheal disease and hunger following the flooding in 1988. Factors involving power, access, location, livelihood, and biology mutually determined a situation of particular unsafe conditions and enhanced vulnerability. (Blaikie et al., 1994: 27)

Across the globe, specific groups of societies are being placed in especially risky circumstances. It has long been recognised that environmental risks and disaster effects are distributed unequally by class, race, gender, ethnicity and age (Alexander, 1993; Blaikie et al., 1994; Cannon, 1994; Hewitt, 1983, 1997; O'Keefe, et al., 1976; Varley, 1994; Wisner et al., 2004). At-risk groups include low-income households, ethnic minorities, old people, and female-headed households, among others. This is often most acutely documented in less developed countries, where development processes can serve to deepen social inequity, but even within more affluent nations, sharp social inequalities persist (Hewitt, 1983, 1997). In more developed nations, marginalised groups are not necessarily confined to areas at greatest risk from environmental hazards. Indeed, it is often the wealthy who choose to live in physically hazardous settings, convinced that it is safe to build palatial homes on hurricane-prone shores, perched precariously on steep unstable slopes or amidst incendiary scrub. They do so, partly because their affluence can buy superior engineering, which affords some degree of protection, but more because the social and economic resilience of the owners offsets their acute physical vulnerability (Wisner et al., 2004).

This uneven distribution of risk derives, in large part, from an institutional landscape in which there is an inequitable distribution of resources – economic, social and political (Blaikie et al., 1994). It is not possible, for example, to gain equal security through employment or subsistence activities; households with

Figure 9.6 These wealthy homes on Mulholland Drive, Los Angeles, have a high physical exposure to landslide and earthquake hazards, but their affluent owners will tend to have a high social and economic resilience to disaster events (photo: the authors)

higher incomes and stable employment have livelihoods that are less vulnerable to hazards. Alongside people's hazard exposure in their housing and work sites, there are cultural traditions and technical knowledge about how and where to live, which sets their capacity to protect themselves. This 'self-protection' is augmented by 'social protection' – the degree of awareness of hazards, warning systems, emergency plans, and regulations to do with building codes and land uses. Inevitably, because state and local-level organisations invariably have those promotional and regulatory obligations, political factors too become implicit elements of vulnerability (Wisner et al., 2004).

Because the propensity to suffer some degree of loss from a damaging and destructive event is registered not by physical exposure to hazards alone, but also by the capacity of a community to withstand those hazards, vulnerability to natural hazards changes through time (Oliver-Smith 1999a; Turner et al., 2003). In a context of chronic vulnerability, catastrophe is probably inevitable. But most communities can adapt to hazardous environs in ways that can reduce their vulnerability, conferring on them a degree of hazard 'resilience' (Berkes, 2007).

Social capital and the building of resilient communities

In a sense, resilience is the flip side of vulnerability – a community that reduces its vulnerability is one with a growing resilience (Buckle et al., 2000). Communities may become more resilient to hazards if they have a range of options for coping with external shocks and stresses. Diversification is a universal strategy aimed at reducing risks by spreading them out, thereby increasing opportunities in the face of hazards (Turner et al., 2003). In social systems, risk can be spread by diversifying the resource base and encouraging alternative lifestyles and activities, assets which are collectively called 'natural capital' (Adger et al., 2005). Thus, a traditionally diverse agro-ecological portfolio offers a more robust buffer to natural hazards than modern monocultures that feed export-driven agriculture (e.g. Holt-Gimenez, 2002). But a diversity of physical assets requires complex social understandings. Among the drought-prone Turkana pastoralists of east Africa, for example, a flexible resource base relies strongly on social organisation, social links and the manipulation of kinship systems (McCabe, 2002).

Many hazard-prone societies have developed adaptations to deal with severe natural disturbances, to the extent that a disaster, at least up to certain extremes, might not even constitute a disaster to them but simply part of their normal experience. Jeffries (1981), for example, notes how the regular hazard-threats endured by villagers on the flanks of Mount Merapi volcano in Indonesia were referred to as *banjir biasa* and *lapar biasa* – literally 'our usual landslide' and 'our regular hunger'. In such societies, traditional practices provide buffers to withstand catastrophe. For instance, rural societies in Bangladesh have adapted agricultural and fish production to periodic floods; in this long-enduring population, the Bangla language differentiates between *bosrha* for 'good' (i.e. normal) flood and *bonna* for 'bad' flood (Haque, 1994). Similarly, South Pacific island communities have been shown to have hazard protocols culturally embedded into their institutions and practices, even though the destructive events in question – tropical storms – recurred every twenty years or so. Adger et al. (2005), however, noted the lack of such a social memory of tsunamis in some areas affected by the 2004 Indian Ocean tsunami where the previous significant wave had struck 60 years before. The implication is that hazard-adapted communities can develop only if disturbances recur sufficiently regularly to allow the appropriate social tools to be maintained.

Resilience is further strengthened by building local and traditional knowledge into hazard education programmes. For instance, Berkes (2007) reported on the potential contributions of Canadian and Alaskan Inuit communities in assessing the impact of global climate changes. Integrating indigenous observations and community-based monitoring into broader hazard studies is important in establishing a dialogue between scientists and those parties directly affected by environmental change. Cronin et al. (2004), for example, used community-participatory methods to integrate local understanding of 'safe' and 'unsafe' areas around volcanic centres in Vanuatu to engender more effective emergency action plans. At the core of this 'bottom-up' approach is the desire to foster educative change, without eroding the cultural values and diversity of the community at risk. It is important to acknowledge, however, that traditional practices may, in some circumstances, make environments more hazardous, and that reliance on indigenous culture is not a panacea for effective risk management (e.g. Morris, 2003).

Finally, the resilience of any system is closely related to its capacity for recovery through self-organisation and self-healing. Although it has long been assumed that governments, from the federal to the municipal, comprise the backbone of emergency management, community organisations have a major role to play in the face of disaster (King, 2007). This community resilience takes the form of networks of strong and weak ties – families, churches, local volunteer and relief groups, hobby clubs, even neighbourhood and crime watch organisations – that is referred to collectively as 'social capital' (Dynes, 2002).

Through social capital, citizens assume roles as active agents rather than passive victims, since they are able to draw upon collective strengths, assistance and resources to deal with hazards and disasters, thereby being more proactive in decision making and effecting a more speedy recovery (e.g. Bolin and Stanford,

Figure 9.7   The fertile volcanic lowlands of central Java, Indonesia, where earthquakes, volcanic eruptions, landslides and famines are part of the ordinary hazard experience (photo: the authors)

1998; Murphy, 2007; Shaw and Goda, 2004). Berkes (2007) cites research in the Pacific, in which the same hurricane striking Samoa and neighbouring American Samoa produced markedly different results: the former were prepared and capable while the latter, much more affluent and used to outside aid for disasters, had weaker institutions for response. What's more, it is important that community organisations link effectively with regional and national organisations. Wisner (2001), for example, argues that so few people died when Hurricane Michelle hit Cuba very hard because of the existence of strong organic links between government and people. Political environments in which there is strong linkage from local to national levels tend to withstand disasters better.

'When all hell breaks loose'

Prior to natural crises, the cultural, social and political factors that promote vulnerability and lessen resilience may be hidden from view. Only after a hazard strikes does the extent to which latent risk factors were present become clear. Hurricanes, floods, fires, earthquakes and volcanic eruptions test the very sinews of society. Within seconds, minutes or hours, the entire fabric of a community, city or region can dissolve. In this context, earthquakes, hurricanes and floods play the role of 'revealers', since it is only 'when all hell breaks loose' that the social and technical resilience of a community is truly tested (Oliver-Smith and Hoffman, 2002). How a community emerges from this maelstrom is unique, specific to the particular circumstances of the tragedy, and it is far from a linear process (King, 2007). Haas et al. (1977) recognised four distinct stages – response, relief, reconstruction and recovery – and argued that each stage took ten times as

Figure 9.8 La Conchita, California, where a mudslide in January 2005 claimed 13 lives. The immediate aftermaths of disasters often see great solidarity within afflicted communities, but later there is often discord with external agencies (photo: the authors)

long as the previous one. If applied to Hurricane Katrina in 2005, the six weeks of emergency response and relief means sixty weeks of reconstruction and almost twelve years of full recovery. Recent studies (e.g. Wisner et al., 2004) have questioned this simple logarithmic model, but, the vagaries of individual disasters accepted, a broadly predictable process of recovery is frequently observed (Hoffman, 1999a).

The immediate aftermath – the response phase – is a time of remarkable energy, action and, often, great social cohesion (Hoffman, 1999a). The depiction of disaster victims as dazed, confused and helpless individuals, suddenly dependent on external handouts, is a fallacy. Instead, the observations across many cultures are that survivors quickly organise into community groups in order to participate in the rescue phase. The initial stage of post-disaster recovery, therefore, is one of unity and community. Solidarity forged in the immediate aftermath of the chaos often blurs previously impenetrable divisions of religion, class, ethnicity, or race.

However, as days turn to weeks and rescue transforms into reconstruction, societal discord can arise (Hoffman, 1999a). A common tendency is for disaster survivors to differentiate themselves from those that simply endured the crisis or witnessed the drama from afar. For example, in the case of the 1991 Oakland (California) firestorm, the criteria about whether or not an individual was seen as a 'disaster victim' became the question of whether or not their *whole* house had been destroyed (Hoffman, 1999a). Pre-disaster norms of inequality, alliance, and allegiance can also reassert themselves as recovery takes hold. After the 1970 Yungay earthquake, for example, as housing was assigned, middle-class survivors did not want to live near Indians in government-built housing (Oliver-Smith,

1999b). Often, the source of the discord relates to the speed or nature of the recovery itself, as the victim groups feel their views on how to rebuild their community conflicts with the ideas and recovery plans of external agencies and authorities. Equally, the conflict may be about aid allocation – with external communities perceiving survivors to be 'getting something for nothing' and their circumstances to be the result of their own negligence or failings (Hoffman, 1999a).

Given the diversity of organisations involved in the aftermath of disasters – local and community groups, media, grassroots organisations, non-governmental organisations, government institutions and departments, emergency services, military and security forces, and international agencies – it is inevitable that problems of mismanagement will emerge (King, 2007). But amid this confusion of ideas, approaches and agendas, there is the potential for deliberate neglect, as power and resources are distributed unevenly, and marginalised groups are left further isolated and weakened (e.g. Murphy, 2007; Pelling, 1998). The result can be what Dyer (2002) calls 'corrosive communities', those whose social capital is so debilitated that there is effectively little opportunity for recovery, and consequently, heightened vulnerability to future hazards. In extreme cases, natural emergencies can motivate instances of cultural enmity and ethnic violence to those 'outside' the community.

The economics of post-disaster recovery

Vibrant and inclusive community partnerships may be a highly effective base from which to resist emergencies turning into disasters, but they go only so far. When an extreme natural event obliterates the physical infrastructure, pre-existing social capital may be overwhelmed (Murphy, 2007). Recovery, therefore, is greatly dependent on the extent to which a region's economic infrastructure was compromised. The loss of critical 'lifelines' that support vital societal functions – electric power, water, transportation and other systems – can result in widespread and prolonged impairment. These critical lifelines are highly interconnected; the disruption to electric power, for instance, is significant not only for its direct impact, but also in triggering or exacerbating disruptions to water, transportation and other systems that in turn cause knock-on societal effects (Chang et al., 2007). The more technologically advanced a nation, the more vulnerable it is to such 'infrastructure failure interdependencies' (Rinaldi et al., 2001).

An example was provided by the 1995 Great Hanshin (Kobe) earthquake. Although this earthquake was moderate in size (M 6.9), it cost over 4000 lives and caused an estimated US$100 billion in damage (Brauner and Cochrane, 1995). The city's capacity to rebuild and recover from this seismic shock was hindered by disruption of its critical infrastructure facilities. While electric power and telecommunications were re-established within the first couple of weeks, and water and natural gas within three to four months, the rail and highway networks were not completely restored until 7 and 21 months respectively, after the earthquakes, and the Port of Kobe required just over two years to complete repairs (Chang, 2000a). The loss of the port facilities was especially crucial: two-thirds of the city's economy – and a sixth of its employment – was tied to the port, which, prior to the disaster, was ranked sixth among container ports worldwide (Chang, 2000b). Extensive damage from liquefaction and ground failure, however, forced the port to shut down, and by the time it became fully operational two years later, much of the container traffic had been diverted to other ports; by 1997, Kobe's global ranking had slumped to 17th (Chang, 2000b).

This permanent downturn in Kobe's post-earthquake fortunes is a reminder that even after critical lifelines are restored and physical damage is repaired, pre-disaster economic conditions may not necessarily be regained (Chang, 2000b). This contrasts with other studies which have suggested that the economic conse- quences of 'natural' disasters are at worst transient, and at best, positive. For one thing, emergency crises may bring economic benefits, such as when a community receives an infusion of disaster relief funds that injects money into the local eco- nomy. For example, the economies of both Miami and Los Angeles appear to have been helped by the 1992 Hurricane Andrew and the 1994 Northridge earthquake respectively (Cochrane, 1997); both economies had been performing poorly before the disasters but rebounded as a result of reconstruction activities. Indeed, Romero and Adams (1995) considered the post-disaster Northridge reconstruction stimulus to have speeded up California's recovery from economic recession. It is in this context that California's earthquakes can be regarded by city planners as positive agents of urban renewal (Holt, 2005); for example, the 1989 Loma Prieta earthquake forced the replacement of a damaged elevated freeway with a pedes- trianised boulevard that has transformed the fortunes of San Francisco's water- front district.

Still, the economic costs of future urban disasters may be truly enormous. The projected costs of Greater Tokyo suffering a repeat of the large (M 7.9) earthquake of 1923, for example, is a deathtoll of 60,000 people and a colossal US$4.3 trillion (Insurance Information Institute, 2004). Such is the severity of the seismic threat to Tokyo that there are even strong pressures to relocate the national cap- ital to a safer site (Kumagai and Nojima, 1999). A direct seismic strike on the metropolis is one of the insurance industry's top five potential event losses, with the most calamitous being an earthquake in the Los Angeles area, where insured losses of $75 billion are anticipated. But ironically, the financial shock waves of the next Tokyo earthquake may reach the USA anyway – Japan holds almost a fifth of US Treasury securities, an amount totalling some US$700 billion (Stein et al., 2006).

The politics of natural disaster

Natural disasters are increasingly becoming international political occasions, either putting governments under pressure if they fail to respond well, or creat- ing new ways for thawing previously icy international relations. The earthquakes that struck Istanbul and Athens, in August and September 1999 respectively, opened communication channels between feuding Greece and Turkey, and the Bam (southern Iran) earthquake of December 2003 prompted offers of aid from forty countries, including the USA – the 'Great Satan' – which had broken off diplomatic relations with Iran 20 years before. Even Hurricane Katrina, whose

Figure 9.9 Conurbations like Greater Tokyo are now major targets for hazardous events that may bring loss of life and economic impacts on a scale that are unprecedented in human history (photo: the authors)

devastation appeared to escalate beyond the American government's ability to cope, triggered offers of aid from nation states with varying degrees of conflict with the USA (Kelman, 2007). Much of the evidence suggests that disaster-related initiatives can significantly influence and reinforce diplomatic processes that have already started, but rarely cement political rapprochement, though a possible exception is the peace deal reached in Aceh after the 26 December 2004 earthquake and tsunami (Kelman, 2006).

Natural hazards may offer opportunities for such 'disaster diplomacy', but they can equally nourish political instability and fuel conflict situations. Pelling and Dill (2006) outline a number of ways in which disasters can trigger political action. These include the tendency for disasters to hit politically peripheral regions hardest, thereby catalysing regional political tension; the potential for post-disaster governmental manipulation to worsen existing regional inequalities; the likelihood of regimes interpreting spontaneous collective actions by afflicted communities in the aftermath of a disaster as a threat, and thereby responding with repression; and the potential for disasters to enhance or even regain the popular legitimacy of political leaders. It has been argued, for example, that the cyclone and storm surge in East Pakistan in 1970 contributed to the development of the Bangladesh independence movement, while the revolutionary movement in Nicaragua from 1974 to 1979 derived some of its impetus from the effect of the Managua earthquake of 1972 (Wisner et al., 2004: 60).

Climate change and land degradation too can inflame regional and ethnic tensions. Crippling droughts and extreme weather have historically contributed to the overthrow of elites and dramatic realignments of power (Davis, 2000), and natural environmental crises are implicated in several 20th-century political crises. Adger (1999), for example, argues that social upheaval, and ultimately peasant rebellion against the colonial French rule in Vietnam during the early 20th century, occurred concurrently with severe droughts and floods, while the 1970 Sahel famine contributed to the toppling of the governments of Niger and Ethiopia for their perceived failings in the crisis (Blaikie et al., 2004: 60). Regardless of whether the ongoing Darfur (Sudan) conflict is rooted in drought (cf. Butler, 2007; UN report, 2007), extreme climate crises can cause food shortages, thereby generating anger against governments and fuelling social unrest. Databases on civil wars and water availability appear to show that when rainfall is significantly below normal, the risk of low-level conflict escalating into full-scale civil war approximately doubles in the following year (Giles, 2007). There is a concern that future global climate change could act as a 'threat multiplier', with events such as droughts toppling unstable governments and unleashing swarms of conflicts.

In this wider geopolitical realm, natural hazards need to be seen as integral elements of the broader issues of environmental security and sustainability (Beer, 2003). Regional and global risk analyses highlight countries and geographical regions with a propensity for natural disasters, and provide institutions, such as the United Nations Development Program (UNDP), the World Bank, and the International Monetary Fund, with an objective basis for drawing attention to nations where disaster management is a priority (Dilley, 2006). In turn, disaster recovery can become the impetus for political and economic reform; after Hurricane Mitch in 1998, for example, afflicted countries in Central America agreed a set of principles with international aid donors that included promotion of democracy and good governance, political decentralisation and economic debt reduction (Wisner et al., 2004). Although there are still few signs of those principles actually being implemented (Wisner et al., 2004), some social commentators warn against using disasters as a pretext for international political and economic 'engineering', a tendency which Klein (2005) calls 'disaster capitalism'.

Calls for change – communicating hazard

Natural calamities can clearly be a bridge to change, occasionally even great change (Hoffman, 1999b). The burst of heightened political awareness following a disaster can often stimulate calls for strategic or institutional action. For example, the avalanche disaster in Switzerland in 1951, which caused 98 fatalities, was the main catalysing factor leading to the construction of over 500 km of structures existing in the Swiss Alps today (Phillips, 2006). Sometimes even the threat of socially unacceptable losses can be a spur to hazard legislation. The destruction of schools in the 1930 Long Beach, 1971 San Fernando and 1994 Northridge earthquakes, for example, ushered in successive changes in Californian seismic building codes (Bolt, 1999). Such popular impetuses would appear to be important in overcoming inherent resistance within state and local communities to increased regulation of often valuable hazard-prone areas (May and Birkland, 1994). But in many hazard-prone regions, even the most basic measures of resilience are not invested in. The 2005 Pakistan earthquake claimed 16,000 children who were attending school at the time the earthquake struck. It is a similar story with critical facilities like hospitals. As many hazard practitioners bitterly point out, the issue is not a technical one. Instead, hazardous events continue to expose the basic societal fallibilities in disaster preparedness and emergency management. In this final section, we examine the role of hazard perception in shaping the actions of at-risk individuals and communities.

Government actions and legislated building standards can build resilience and reduce vulnerability, but individuals too can act to limit damage caused by natural hazards. In known hazardous areas, substantial funds are expended annually on risk communication programmes to promote preparedness (e.g. storing food and water, securing household items and furniture to prevent injury, and preparing a household evacuation plan). When citizens take steps to ensure that their homes meet or exceed the relevant building standards, and that they have provisions for survival after a disaster, losses may be substantially reduced (Kreps, 1984; Sorenson and Mileti, 1987). And yet, a common finding in research on natural disasters is that people living in at-risk communities continue to demonstrate poor knowledge of risk-mitigation procedures and a reticence to adopt protective measures (Johnston et al., 1999; Paton and Johnston, 2001).

Endemic unpreparedness for hazard threats has been widely documented for earthquakes. Studies in California, for instance, have found that very few people surveyed had made any structural changes to their home, purchased earthquake insurance (Jackson, 1981) or had appropriate first-aid supplies (Turner et al., 1986), and a similar dearth of basic precautionary care was evident among earthquake-vulnerable communities in Turkey (Rustlemi and Karanci, 1999). Although a link between perceived risk and actual geophysical risk might be logically assumed, this may not always be justified (e.g. Palm and Hodgson, 1992). For example, residents of Los Angeles surveyed after an earthquake in 1989 readily acknowledged the threat of earthquakes but had a low perception of personal risk (Burger and Palmer, 1992). A prominent reason for this is that it is not information about hazards per se that determines action, but how people perceive that information socially and culturally (Paton et al., 2000). To be effective, risk communication efforts need to acknowledge these socio-cultural vagaries; passive presentation of hazard information is unlikely to produce anticipated changes in risk behaviour (McIvor and Paton, 2007). In an evaluation of a volcanic risk-communication programme, for example, Ballantyne et al. (2000) reveal that provision of hazard information resulted in over a quarter of the respondents feeling *less* concerned about hazards. Adams (1995) refers to this tendency as 'risk homeostasis', whereby a perceived increase in safety (e.g. new safety practices, training, etc.) can reduce the risk attributed to a hazard, increase risky behaviour and render individuals more vulnerable to hazardous incidents.

In part, this skewed appreciation of natural risk is because, left to individual choices, most people display an unrealistic 'optimistic bias' – a pattern of judgements where people see themselves less likely to be harmed by future risks than others. Optimistic bias emerges in response to a wide range of risk-taking activities – from motorcycle use and bungee jumping to exposure to health risks such as radon – but it is frequently linked to people's experiences of natural disaster (Spittal et al., 2005 and references therein). A common trait in hazard-prone environments is that while many people accept they will experience a hurricane, flood or earthquake, most believe that it would not harm them or their property. Even after having lived through a damaging event, an individual's optimistic bias re-emerges (Burger and Palmer, 1992). In fact, first-hand experience may reduce the uptake of preventative measures, if people infer from their prior ability to deal with minimally disruptive hazard events a capacity to deal with any future occurrence (Johnston et al., 1999; Paton et al., 1998).

Much public-hazard education makes the erroneous assumption that natural hazards are equally as significant to people as social hazards, such as crime and unemployment, and that people can evaluate their salience independently (Paton, 2003). But it turns out that sources of adversity encountered on a daily basis, or those whose existence and implications are reiterated through regular media attention, are perceived as being most important. Thus, in the extended lulls between hazard events, when most readiness work must take place, natural hazards are out-competed by their social counterparts for an individual's attention (Miller et al., 1999). Only when natural threats become prominent, such as in the period before landfall of a tropical storm or in the precursory phase of volcanic activity, do hazards become salient – in other words, they become a significant social talking point – and a motivating force for protective behaviour.

It appears that the more people engage in discourse about hazard issues (and seek advice from those deemed knowledgeable about them), the more salient these issues will be perceived by a community, and the greater the likelihood of protective measures (McIvor and Paton, 2007). Because an individual's level of resilience can be correlated with their level of involvement in community activities (membership of clubs, social action groups, etc.), the more people engage in activities that engender a sense of community, the greater will be their resilience to adversity (Paton and Johnston, 2001). What is emerging from such studies is a realisation that public-awareness programmes currently targeted at individuals might be more usefully directed towards groups (e.g. churches, social clubs and neighbourhood associations) and community development initiatives that build and bolster social capital (Quarantelli, cited in Wisner et al., 2004).

Summary

Natural hazards are still often celebrated for being exceptional, rather than systemic, elements of the environments in which we dwell. As Hewitt (1983: 25) reminds us, 'in most places and segments of society where calamities are occurring, the natural events are about as certain as anything within a person's lifetime'. And yet, three decades after hazard practitioners reassigned them to the social, economic, cultural and political milieu of ordinary, everyday life, scientists, governments and even international institutions are still seeking scientific and engineered solutions to problems that are inextricably rooted in complex societal systems. The message in this chapter is simple: hazards are natural; disasters need not be.

Case study 1: Post-Katrina New Orleans – smaller and whiter?

Hurricane Katrina was not the strongest of the three storms that reached category 5 intensity during the 2005 Atlantic hurricane season, but its strength on landfall near New Orleans made it the most devastating disaster in US history. An area the size of the United Kingdom was involved, displacing more than one million people, claiming over 1,000 lives, and exceeding $80 billion in costs (Cutter et al., 2006). And yet not only was this event foreseen (e.g. Fischetti, 2001), but forecasts of the storm track were accurate and gave three days' warning to authorities in New Orleans (McCallum and Heming, 2006). What surprised few was that the ageing infrastructure of the Mississippi coast's flood protection levees – designed for a category 3 storm surge – failed under the onslaught of the storm, allowing widespread inundation of the city. What stunned many was the resulting institutional meltdown, which for several days left evacuees with no power, no drinking water, dwindling food supplies, understaffed law enforcement, and delayed search and rescue activities (Cutter et al., 2006).

At first glance, Hurricane Katrina seemed to have demolished neighbourhoods without paying heed to the colour, creed, or status of their residents. Multi-million dollar beach homes and floating casinos were destroyed by equivalent winds and storm surges to those that battered low-income residential areas. But while affluent coastal strips and suburbs had been largely evacuated, many inner-city residents remained. In a city 87 per cent black and 30 per cent poor, there were 112,000 households without private vehicles, and because Katrina struck on 29 August – two days before pay cheques and welfare or disability cheques would arrive – many had no money for transportation (Cutter et al., 2006).

Soon after the chaos of evacuation, many of the Gulf Coast's more affluent residents and the casinos along the shoreline had received insurance settlements and were already rebuilding, content to live with the risks (Cutter et al., 2006). It was less clear how many would return to low-income neighbourhoods clogged with debris and damaged buildings. If people returned only to undamaged neighbourhoods, it was reckoned that New Orleans would lose 80 per cent of its black residents but only 50 per cent of its white ones (Biever, 2006). In fact, many are returning to shelter in damaged homes; in doing so, they are facing the prospects of unaffordable housing and low-paid service jobs, but of those relocated many

Figure 9.10   Beachfront properties remain an attractive opportunity for the affluent, even along hazard-prone coasts (photo: the authors)

have gone permanently. It remains to be seen whether a large influx of Hispanic workers, drawn by vacated service-sector jobs and new employment in recon-struction, produces a very different racial mix, but on the face of it, New Orleans looks set to emerge from the wreckage of Katrina smaller and whiter (Biever, 2006).

Case study 2: The Asian tsunami

The great Sumatran (Indonesia) earthquake of 26 December 2004 that radiated tsunami destruction around the Indian Ocean was a truly global disaster. The event cost 250,000 lives and huge economic losses, but its reach extended far beyond the countries in the devastated region, killing large numbers of tourists from distant affluent nations; for example, it resulted in the greatest loss of life of Swedish citizens from a natural event (Huppert and Sparks, 2006). With the threat of large tsunami-generating earthquakes known before the Boxing Day disaster (Sieh, 2006), its aftermath was dominated by international calls for a tsunami warning system akin to that protecting the Pacific Ocean (Basher, 2006). However, the lack of emergency preparedness and very short lead-time for tsunami evacu-ation along much of Indonesia's earthquake coast convinces many that the answer to future tsunami mitigation lies not in high-tech early warning systems or even the hugely expensive construction of sea defences (Sieh, 2006). Instead, the fact that some fishing communities survived the tsunami, thanks to inherited local knowledge and to institutional preparedness for disasters (Adger et al., 2005), supports the view that educational outreach programmes, coupled with improved construction and land-use planning, is a more effective means of build-ing tsunami-resilient communities (Sieh, 2006) (for examples of Pacific-based pro-grams, see Dengler, 2005 and Jonientz-Trisler et al., 2005).

Part of that resilience may come from the coastal environment itself. On some distant coasts, the brunt of the tsunami's impact was borne by coral reefs, coastal dunes and mangrove swamps (e.g. Kesavan and Swaminathan, 2006), though in many places these natural barriers had been removed for tourist developments. In Banda Aceh, Indonesia, however, the presence of such coastal buffers made no difference to the impact of the devastating waves. Nevertheless, throughout south-east Asian shores, deforestation of mangroves for intensive shrimp farming, a lucrative export industry, has reduced livelihood options for local farming and fishing communities (Keys et al., 2006). This, together with other forms of chronic degradation (land clearance, coastal erosion, overfishing and coral mining) has significantly reduced the potential for economic recovery from the tsunami, because of the loss of traditional income sources related to diverse coastal ecosys-tems. Attempts to rebuild coral reefs damaged by the tsunami are ineffective in the face of their chronic degradation by pollution, over-fishing and destructive fishing practices. Adger et al. (2005: 1038) argue that rather than squander funds on simplistic coral-reef rehabilitation projects, 'support should be directed to pro-vide ecologically sustainable, long-term employment for coastal communities to eliminate poverty, and to improve the local and regional governance systems for

Figure 9.11   Devastated tourist complexes at Phuket, Thailand, a few weeks after the 26 December 2004 tsunami (photo: the authors)

managing the natural resilience of coral reefs.' However, deep systemic economic problems still exist within many of these nations, and for many of them it is tourism that is seen as the way out. Less than four weeks after the tsunami, the head of Thailand's tourism authority implored lucrative Western tourists to come back with the words (cited in Keys et al., 2006): 'Although we have lost many lives and much property in the disaster, the tragic event has brought some good things, as it has swept away all the garbage and some parts of the Andaman Sea around Phuket are the cleanest they've been in twenty years.'

References

Adams, J. (1995) *Risk*. London: UCL Press.

Adger, W.N. (1999) 'Evolution of economy and environment: an application to land use in lowland Vietnam', *Ecological Economics*, 31: 365–79.

Adger, W.N. and Brooks, N. (2003) 'Does global environmental change cause vulnerability to disaster?', in M. Pelling (ed.), *Natural Disasters and Development in a Globalising World*. London: Routledge. pp. 19–42.

Adger, W.N., Hughes, T.P., Folke, C., Carpenter, S.R. and Rockstrom, J. (2005) 'Social-ecological resilience to coastal distasters', *Science*, 309: 1036–9.

Alexander, D. (1993) *Natural Disasters*, London: UCL Press.

Ali, A.M.S. (2007) 'September 2004 flood event in southwestern Bangladesh: a study of the nature, causes and human perception and adjustments to a new hazard', *Natural Hazards*, 40: 89–111.

Ballantyne, M., Paton, D., Johnston, D., Kozuch, M. and Daly, M. (2000) 'Information on volcanic and earthquake hazards: the impact on awareness and preparation. Wellington: Institute of Geological and Nuclear Sciences, Report No. 2.

Basher, R. (2006) 'Global early warning systems for natural hazards: systemic and people-centred', *Philosophical Transactions of the Royal Society A*, 364: 2167–82.

Beer, T. (2003) 'Environmental risk and sustainability', in T. Beer and A. Ismail-Zadeh (eds), *Risk Science and Sustainability*, Dordrecht: Kluwer Academic. pp. 39–61.

Berkes, F. (2007) 'Understanding uncertainty and reducing vulnerability: lessons from resilience thinking', *Natural Hazards*, 41: 283–95.

Biever, C. (2006) 'New Orleans fights for survival', *New Scientist*, 28 February.

Blaikie, P., Cannon, T., Davies, I. and Wisner, B. (1994) *At Risk: Natural hazards, People's Vulnerability and Disasters*. 1st edition. London: Routledge.

Bolin, R. and Stanford, L. (1998) 'The Northridge earthquake: community-based approaches to unmet recovery needs', *Disasters*, 22: 21–38.

Bolt, B.A. (1999) *Earthquakes*. 4th edition. New York: W.H. Freeman.

Brauner, C. and Cochrane, S. (1995) *The Great Hanshin Earthquake, Kobe: Trial, Error, Success*. Zurich: Swiss Reinsurance Company.

Buckle, P., Mars, G. and Smale, S. (2000) 'New approaches to assessing vulnerability and resilience', *Australian Journal of Emergency Management*, Winter: 8–14.

Burger, J.M. and Palmer, M.I. (1992) 'Changes in and generalization of unrealistic optimism following experiences with stressful events: reactions to the 1989 California earthquake', *Journal of Personality and Social Psychology*, 18: 39–43.

Burton, I., Kates, R.W. and White, G.F. (1978) *The Environment as Hazard*. Oxford: Oxford University Press.

Butler, D. (2007) 'Darfur's climate roots challenged', *Nature*, 447 (7148): 1038.

Cannon, T. (1994) 'Vulnerability analysis and explanation of "natural" disasters', in A. Varley (ed.), *Disasters, Development and Environment*. Chichester: Wiley. pp. 13–30.

Cardona, O.D. (2005) Indicators of disaster risk and risk management: program for Latin America and the Caribbean. Summary Report. Washington, DC: Inter-American Development Bank.

Chang, S.E. (2000a) 'Transportation performance, disaster vulnerability and long-term effects of earthquakes', Second Euroconference on Global Change and Catastrophe Risk Management, Laxenberg, Austria, 6–9 July 2000. http://www.iiasa.ac.at/Research/RMP/july2000/papers.html

Chang, S.E. (2000b) 'Disasters and transport systems: loss of recovery and competition at the Port of Kobe after the 1995 earthquake', *Journal of Transport Geography*, 8: 53–65.

Chang, S.E., McDaniels, T.L., Mikawoz, J. and Peterson, K. (2007) 'Infrastructure failure interdependencies in extreme events: power outage consequences in the 1998 Ice Storm', *Natural Hazards*, 41: 337–58.

Cochrane, H.C. (1997) 'Forecasting the economic impact of a Mid-West earthquake', in B.G. Jones (ed.), *Economic Consequences of Earthquakes: Preparing for the Unexpected*. New York: New York Center for Earthquake Engineering Research. pp. 223–48.

Cronin, S.J., Gaylord, D.R., Charley, D., Alloway, B.V., Wallez, S. and Esau, J.W. (2004) 'Participatory methods of incorporating scientific with traditional knowledge for volcanic hazard management on Ambae Island', *Bulletin of Volcanology*, 66: 652–88.

Cutter, S.L., Emrich, C.T., Mitchell, J.T., Boruff, B.J., Schmidtlein, M.T., Burton, C.G. and Melton, G. (2006) 'The long road home: race, class and recovery from Hurricane Katrina', *Environment*, 48: 9–20.

Davis, M. (2000) *Late Holocene Holocausts*. London: Verso.

Dengler, L. (2005) 'The role of education in the National Tsunami Hazard Mitigation Program', *Natural Hazards*, 35: 141–53.

UNDERSTANDING ENVIRONMENTAL ISSUES

Dilley, M., Chen, R.S., Deichmann, U., Lerner-Larn, A.L. and Arnold, M. (2005) 'Natural disaster hotspots: a global risk analysis.' Washington, D.C.: International Bank for Reconstruction and Development/The World Bank and Columbia University.

Dilley, M. (2006) 'Setting priorities: global patterns of disaster risk', *Philosophical Transactions of the Royal Society A*, 364: 2217–29.

Downton, M.W. and Pielke, Jr., R.A. (2005) 'How accurate are disaster loss data? The case of US flood damage', *Natural Hazards*, 35: 211–28.

Dyer, C.L. (2002) 'Punctuated entropy as culture-induced change: the case of the Exxon Valdez oil spill', in S.M. Hoffman and A. Oliver-Smith (eds), *Catastrophe and Culture: The Anthropology of Disaster*. Santa Fe, NM: School of American Research Press. pp. 159–86.

Dynes, R.R. (2002) 'The importance of social capital in disaster response'. Preliminary Paper No. 327. Delaware, MD: University of Delaware, Disaster Research Centre.

Eckholm, E.P. (1975) 'The deterioration of mountain environments', *Science*, 189: 764–70.

EM-DAT (2006) 'The OFDA/CRED International Disaster Database'. Brussels: Université Catholique de Louvain. (www.cred.be/emdat)

Fengqing, J., Cheng, Z., Guijin, M., Ruji, H. and Qingxia, M. (2005) 'Magnification of flood disasters and its relation to regional precipitation and local human activities since the 1980s in Xinjiang, Northwestern China', *Natural Hazards*, 36: 307–30.

Fischetti, M. (2001) 'Drowning in New Orleans', *Scientific American*, October: 77–85.

Gardner, J.S. (2003) 'Natural hazards risk in the Kullu district, Himachal Pradesh, India', *The Geographical Review*, 92: 282–306.

Giles, J. (2007) 'Rainfall records could warn of war', *New Scientist*, 2 June: 12.

Greenough, G., McGeehin, M., Bernard, S.M., Trtanj, J., Riad, J. and Engelberg, D. (2001) 'The potential impacts of climate variability and change on health impacts on extreme events in the United States', *Environmental Health Perspectives*, 109: 191–8.

Guha-Sapir, D., Hargitt, D., and Hoyois, Ph. (2004) *Thirty Years of Natural Disasters 1974–2003: The Numbers*. Louvain-la Neuve: Presses Universitaires de Louvain.

Haas, E., Kates, R. and Bowden, M. (1977) *Reconstruction Following Disaster*. Cambridge, MA: MIT Press.

Haque, C.E. (1994) *Hazards in a Fickle Environment: Bangladesh*. Dordrecht: Kluwer.

Hewitt, K. (ed.) (1983) *Interpretations of Calamity*. Winchester: Allen and Unwin.

Hewitt, K. (1997) *Regions of Risk: A Geographical Introduction to Disasters*. Harlow: Longman.

Hoffman, S.M. (1999a) 'The worst of times, the best of times: towards a model of cultural response to disaster', in A. Oliver-Smith and S.M. Hoffman (eds), *The Angry Earth: Disaster in Anthropological Perspective*. London: Routledge. pp. 134–55.

Hoffman, S.M. (1999b) 'After Atlas shrugs: cultural change or persistence after disaster', in A. Oliver-Smith and S.M. Hoffman (eds), *The Angry Earth: Disaster in Anthropological Perspective*. London: Routledge. pp. 302–25.

Holt, T. (2005) 'Let's rubble – longing for the next Big One', *San Francisco Chronicle*, 27 November, E1–E6.

Holt-Gimenez, E. (2002) 'Measuring farmers' agroecological resistance after Hurricane Mitch in Nicaragua: a case study in participatory, sustainable land management impact monitoring', *Agriculture, Ecosystems and Environment*, 93(1): 87–105.

Huppert, H.G. and Sparks, R.S.J. (2006) 'Extreme global hazards: population growth, globalisation and environmental change', *Philosophical Transactions of the Royal Society A*, 364: 1875–88.

Insurance Information Institute (2004) 'Catastrophes: insurance issues', *Hot topics and issues updates*, November (http://www.iii.org/media)

IPCC (2001) *Climate Change: The Scientific Basis*. WMO/UNEP.

ISDR (2004) 'The international strategy for disaster reduction terminology: basic terms of disaster risk reduction, www.unisdr.org

Jackson, E.L. (1981) 'Response to earthquake hazard: the west coast of North America', *Environment and Behavior*, 14: 387–416.

Jeffries, S.E. (1981) 'Our usual landslide: ubiquitous hazard and socio-economic causes of natural disasters in Indonesia', Natural Hazards Research Working Paper No. 40. Boulder: University of Colorado.

Johnston, D.M., Bebbington, M.S., Lai, C.D., Houghton, B.F. and Paton, D. (1999) 'Volcanic hazard perceptions: comparative shifts in knowledge and risk', *Disaster Prevention and Management*, 8: 118–27.

Jonientz-Trisler, C., Simmons, R.S., Yanagi, B.S., Crawford, G.L., Darienzo, M., Eisner, R.K., Petty, E. and Priest, G.R. (2005) 'Planning for tsunami-resilient communities', *Natural Hazards*, 35: 121–39.

Kelman, I. (2006) 'Acting on disaster diplomacy', *Journal of International Affairs*, 59: 215–40.

Kelman, I. (2007) 'Hurricane Katrina disaster diplomacy', *Disasters*, in press.

Kesavan, P.C. and Swaminathan, M.S. (2006) 'Managing extreme natural disasters in coastal areas', *Philosophical Transactions of the Royal Society A*, 364: 2191–216.

Keys, A., Masterman-Smith, H. and Cottle, D. (2006) 'The political economy of a natural disaster: the Boxing Day Tsunami, 2004', *Antipode*, 38: 195–204.

King, D. (2007) 'Organisations in disasters', *Natural Hazards*, 40: 657–65.

Klein, N. (2005) 'The rise of disaster capitalism', *The Nation*, 2 May.

Kreps, G.A. (1984) 'Sociological inquiry and disaster research', *Annual Review of Sociology*, 10: 309–30.

Kumagai, Y. and Nojima, Y. (1999) 'Urbanization and disaster mitigation in Tokyo', in J.K. Mitchell (ed.), *Crucibles of Hazard: Mega-cities and Disasters in Transition*. New York: United Nations University Press. pp. 56–91.

May, P.J. and Birkland, T.A. (1994) 'Earthquake risk reduction: an examination of local regulatory efforts', *Environmental Management*, 18: 923–37.

McCabe, J.T. (2002) 'Impact of drought among Turkana pastoralists: implications for anthropogical theory and hazards research', in S.M. Hoffman and A. Oliver-Smith (eds), *Catastrophe and Culture: The Anthropology of Disaster*. Santa Fe: School of American Research Press. pp. 213–36.

McCallum, E. and Heming, J. (2006) 'Hurricane Katrina – an environmental perspective', *Philosophical Transactions of the Royal Society A*, 364: 2099–115.

McIvor, D. and Paton, D. (2007) 'Preparing for natural hazards: normative and attitudinal influences', *Disaster Prevention and Management*, 16: 79–88.

Miller, M., Paton, D. and Johnston, D. (1999) 'Community vulnerability to volcanic hazard consequences', *Disaster Prevention and Management*, 4: 255–60.

Mitchell, J.F.B., Lowe, L.A., Wood, R.A. and Vellinga, M. (2006) 'Extreme events due to human-induced climate change', *Philosophical Transactions of the Royal Society A*, 364: 2117–33.

Morris, A. (2003) 'Understandings of catastrophe: the landslide at La Josephina, Ecuador', in M. Pelling (ed.), *Natural Disaster and Development in a Globalizing World*. Abingdon: Taylor and Francis. pp. 157–69.

Moseley, M.E. (1999) 'Convergent catastrophe: past patterns and future implications of collateral natural disasters in the Andes', in A. Oliver-Smith and S.M. Hoffman (eds), *The Angry Earth: Disaster in Anthropological Perspective*. London: Routledge. pp. 59–71.

Moseley, M.E. (2002) 'Modelling protracted drought, collateral natural disaster and human responses in the Andes', in S.M. Hoffman and A. Oliver-Smith (eds), *Catastrophe and Culture: the Anthropology of Disaster*. Santa Fe: School of American Research Press. pp. 213–36.

MunichRe (2006) *Hurricanes – More Intense, more Frequent, more Expensive: Insurance in a Time of Changing Risk*. Munich: Münchener Rückversicherungs-Gesellschaft.

Murphy, B.L. (2007) 'Locating social capital in resilient community-level emergency management', *Natural Hazards*, 41: 297–315.

Myers, N. (1986) 'Environmental repercussions of deforestation in the Himalaya', *Journal of World Forestry Resource Management*, 2: 63–72.

Nelson, F.E., Anisimov, O.A. and Shiklomanov, N.I. (2002) 'Climate change and hazard zonation in the circum-Arctic permafrost regions', *Natural Hazards*, 26: 203–25.

O'Keefe, P., Westgate, K. and Wisner, B. (1976) 'Taking the naturalness out of natural disasters', *Nature*, 260 (15 April): 566–7.

Oliver-Smith, A. (1999a) 'What is a disaster?: anthropological perspectives on a persistent question', in A. Oliver-Smith and S.M. Hoffman (eds), *The Angry Earth: Disaster in Anthropological Perspective*. London: Routledge. pp. 18–33.

Oliver-Smith, A. (1999b) 'Peru's five-hundred year earthquake: vulnerability in historical context', in A. Oliver-Smith and S.M. Hoffman (eds), *The Angry Earth: Disaster in Anthropological Perspective*. London: Routledge. pp. 74–88.

Oliver-Smith, A. and Hoffman, S.M. (2002) 'Why anthropologists should study disasters', in S.M. Hoffman and A. Oliver-Smith (eds), *Catastrophe and Culture: The Anthropology of Disaster*. Santa Fe: School of American Research Press. pp. 3–21.

Palm, R. and Hodgson, M.E. (1992) 'After a California earthquake: attitude and behavior change. Geography Research Paper, 233, University of Chicago.

Paton, D. (2003) 'Disaster preparedness: a social-cognitive perspective', *Disaster Prevention and Management*, 12: 210–16.

Paton, D. and Johnston, D. (2001) 'Disasters and communities: vulnerability, resilience and preparedness', *Disaster Prevention and Management*, 10: 270–7.

Paton, D., Johnston, D. and Houghton, B. (1998) 'Organisational responses to a volcanic eruption', *Disaster Prevention and Management*, 7: 5–13.

Paton, D., Smith, L. and Violanti, J. (2000) 'Disaster response: risk, vulnerability and resilience', *Disaster Prevention and Management*, 9: 173–80.

Pelling, M. (1998) 'Participation, social capital and vulnerability to urban flooding in Guyana', *Journal of International Development*, 10: 469–86.

Pelling, M.A. (2002) *The Vulnerability of Cities: Natural Disasters and Social Resilience*. London: Earthscan.

Pelling, M. (ed.) (2003) *Natural Disaster and Development in a Globalizing World*. Abingdon: Taylor & Francis.

Pelling, M. and Dill, K. (2006) '"Natural" disasters as catalysts of political action'. ISP/NSC Briefing Paper 06/01, Chatham House. pp. 4–6.

Pérez-Maqueo, O., Intralawan, A., and Martínez, M.L. (2007) 'Coastal disasters from the perspective of ecological economics', *Ecological Economics*, 63: 273–84.

Phillips, M. (2006) 'Avalanche defence strategies and monitoring of two sites in mountain permafrost terrain, Pontresina, eastern Swiss Alps', *Natural Hazards*, 39: 353–79.

Press, F. and Hamilton, R.M. (1999) 'Mitigating natural disasters', *Science*, 284: 1987.

Rinaldi, S.M., Peerenboom, J.P. and Kelly, T.K. (2001) 'Critical infrastructure dependencies', *IEEE Control Systems Magazine*, December: 11–25.

Romero, T.J. and Adams, J.L. (1995) 'Economic impact of the Northridge earthquake', in M.C. Woods and W.R. Seiple (eds), *The Northridge, California Earthquake of 17 January 1994.* California Department of Conservation, Division of Mines and Geology. Spec. Publ. 116: 263–71.

Rustlemi, A. and Karanci, A.N. (1999) 'Correlates of earthquake cognitions and preparedness behavior in a victimised population', *The Journal of Social Psychology*, 139: 91–101.

Shaw, R. and Goda, K. (2004) 'From disaster to sustainable civil society: the Kobe experience', *Disasters*, 28: 16–40.

Sieh, K. (2006) 'Sumatran megathrust earthquakes: from science to saving lives', *Philosophical Transactions of the Royal Society A*, 364: 1947–63.

Smith, K. (2001) *Environmental Hazards: Assessing Risk and Reducing Disaster*, 3rd edition. London: Routledge.

Smolka, A. (2006) 'Natural disasters and the challenge of extreme events: risk management from an insurance perspective', *Philosophical Transactions of the Royal Society A*, 364: 2147–65.

Sorenson, J.H. and Mileti, D. (1987) 'Programs that encourage the adoption of precautions against natural hazards: review and evaluation', in N. Weinstein (ed.), *Taking Care: Understanding and Encouraging Self-protective Behaviour*. New York, NY: Cambridge University Press.

Spittal, M.J., McClure, J., Siegart, R.J. and Walkey, F.H. (2005) 'Optimistic bias in relation to preparedness for earthquakes', *The Australian Journal of Disaster and Trauma Studies*, 2005(1).

Stein, R.S., Toda, S., Parsons, T. and Grunewald, E. (2006) 'A new probabilistic seismic hazard assessment for greater Tokyo', *Philosophical Transactions of the Royal Society A*, 364: 1965–98.

Trenberth, K. (2005) 'Uncertainty in hurricanes and global warming', *Science*, 308: 1753.

Turner, R.H., Nigg, J.M. and Paz, D.H. (1986) *Waiting for Disaster: Earthquake Watch in California*, Los Angeles: University of California Press.

Turner, B.L. II, Kasperson, R.E., Matson, P.A. et al. (2003) 'A framework for vulnerability analysis in sustainability science.' *Proceedings of the National Academy of Sciences USA*, 100: 8074–9.

Twigg, J. (1998) 'Disasters, development and vulnerability', in J. Twigg (ed.), *Development at Risk?* Report for the UK National Coordination Committee for the International Decade for Natural Disaster Reduction (IDNDR).

UN Environment Programme (2007) *Sudan: Post-conflict Environmental Assessment*. New York. p. 358.

Van der Vink, G., Allen, R.M., Chapin, J., Crooks, M., Fraley, W., Krantz, J., Lavigne, A.M., LeCuyer, A., MacColl, E.K., Morgan, W.J., Ries, B., Robinson, E., Rodriquez, K., Smith, M. and Sponberg, K. (1998) 'Why the United States is becoming more vulnerable to natural disasters', *EOS Transactions*, 79: 533–7.

Varley, A. (1994) (ed.) *Disasters, Development and Environment*. Chichester: Wiley.

Wang, H.B., Sassa, K. and Xu, W.Y. (2006) 'Analysis of a spatial distribution of landslides triggered by the 2004 Chuetsu earthquakes of Niigata Prefecture, Japan', *Natural Hazards*, 41: 43–60.

Wisner, B. (2001) 'Lessons from Cuba? Hurricane Michelle, November', Radix, Radical Interpretations of Disasters. http://online: northumbrian.ac.uk/geography_research/radix/cuba:html

Wisner, B. (2003) 'Floods and mudslides in Algiers: Why no warning? Why poor drainage? Why?', http://www.radixonline.org/algeria.htm

Wisner, B., Blaikie, P., Cannon, T. and Davis, I. (2004) *At Risk: Natural Hazards, People's Vulnerability and Disasters*, 2nd edition. Abingdon: Routledge.

10 Mexico City

Mike Turner

Learning outcomes

Knowledge and understanding of:
- where Mexico City is
- the main physical characteristics of the city and its surrounding area (The Basin of Mexico)
- some demographic characteristics of the city. Some socio-economic characteristics of the city
- the nature of the main environmental problems of the city
- the demographic, social and economic processes which have led to the present environmental situation
- the complexity of the physical, social, economic and, particularly, political difficulties in addressing the environmental situation.

Critical awareness and evaluation:
- you will have learnt that the costs of environmental problems are unevenly distributed among social groups and be able to make judgements about them
- you will have become aware that there is no simple answer to the problems and that, among the different attempted and proposed solutions, there will be gainers and losers – locally, nationally and internationally
- you will be able to make judgements about the various solutions, actual and proposed, and be able to evaluate them in relation to broader political and environmental principles

Finally, you will be able to relate the contents of this chapter to the rest of the book. It is intended as a case study and, as such, it should exemplify, in a single city, many of the issues identified in earlier chapters.

Introduction

> The urban ills that plague Mexico City may be so far advanced as to preclude any viable rescue attempts. An ominous prospect for a megalopolis that is paradigmatic of megacities throughout the developing world.
>
> (Kasperson et al., 1999: xv)

Quotations at the beginning of chapters are often included as a catchy way of starting off. This one however has a more serious purpose – to explain why Mexico City has been chosen as a case study. As the quotation suggests, Mexico City is an extreme case, but it may well be representative of the way that a large number of huge cities throughout the developing world are heading. The details will, of course, differ from place to place but the principal forces underlying a possible descent into catastrophe (and hopes for averting it), which are examined in this chapter, are to be found in all such cities.

In urban areas, the relationships between the environment and the city are extremely complex, but one way of simplifying them a little can be to distinguish between:

> Cities as a threat to the environment
> The environment as a threat to cities
> Social processes as mediators of environmental impacts and costs
>
> (Hall, 2006: 154–5)

However, it will be clear that these are, in fact, *inter*relationships which render it impossible to fully identify 'one-way' effects. Nonetheless, although these divisions are not very helpful when trying to describe and explain reality, they do help to form a general conceptual background when trying to explain very complex situations. This is nowhere better exemplified than in Mexico City.

We will consider first the physical environment, which has unique characteristics leading to its effects on the city being both profound and varied.

Location and physical characteristics

Location

The most important thing to note from Figure 10.1 is that Mexico City is in the tropics and in the mountainous central south of Mexico. These two locational characteristics have substantial environmental effects.

The physical setting

Mexico City is located within an inland drainage basin and, in order to examine the physical environment, it is more appropriate to consider the Basin of Mexico as a whole (Ezcurra et al., 1999), rather than just the urban limits of the city, because it is the basin which forms the local environment of the city.

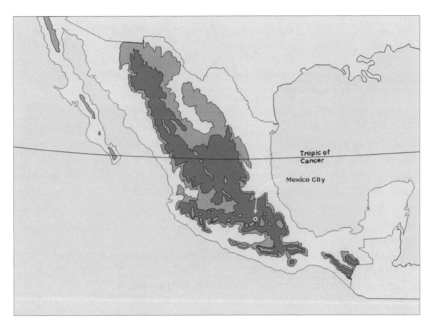

Figure 10.1   Mexico City: location
Source: author

Figure 10.2 shows the city ringed by mountains which form the Basin of Mexico. There is no escape to the sea for rivers, so they all flow towards the centre. Also, the floor of the basin is at an altitude of around 2250 metres which, in association with other factors, itself causes problems, as we shall see in the section on air quality.

Figure 10.3 is a picture of the ancient Aztec capital, Tenochtitlán, on top of which Mexico City was subsequently built. The first thing to notice is that the city is in the middle of a lake. The lake, of course, is a salt lake, because it is in an inland basin. Even at that time (from about 1325 to 1520), the Aztecs had difficulty with water supply and disposal (though they dealt with them very successfully). These problems remain to this day and for the same reasons. Also, the lake or, more specifically, the lake bed is a major element in several of the environmental problems facing the city today.

Although the basin is not arid (it is semi-arid in the north-east but much less so in the west), the watershed is not really big enough, and the rainfall not high enough, to sustain rivers which can provide for the needs of the city. Also, wastewater disposal is difficult because there is no natural gradient to take water away from the basin. These are problems which will be considered later.

In the background of the picture there are two big volcanoes. They can also be seen in Figure 10.2, in the south-east corner of the basin. These are Mexico's famous active volcanoes, Popocatépetl and Iztaccíhuatl, both over 5000 metres high (Popo is the one to the south, to the right in Figure 10.3). Also in Figure 10.2, the cones of several extinct volcanoes can be identified within the basin. The presence of volcanoes, past and present, suggests an area of tectonic instability.

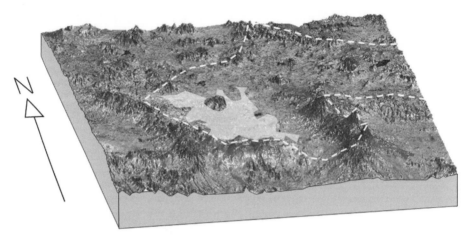

Figure 10.2   Three-dimensional satellite view of the Basin of Mexico. The approximate area of the city is shown in light grey and the black dashed line is the approximate boundary of the basin
Source: author. Satellite data courtesy of the University of Maryland

Figure 10.3   Tenochtitlán
Source: unknown

Additionally, much of the present city, including the central area, is built on the lake bed. In an area of crustal instability such as this, the unconsolidated lake bed greatly exacerbates the effects of any earth movement.

One thing that cannot be seen directly from the satellite image, but which can be deduced if one knows anything of meteorology, is that a basin surrounded by high mountains is susceptible to temperature inversions. In these conditions, cold air is trapped in the basin which would normally mean that it is just extra cold. However, any pollutant emissions are themselves trapped in the cold air, thus

(a)

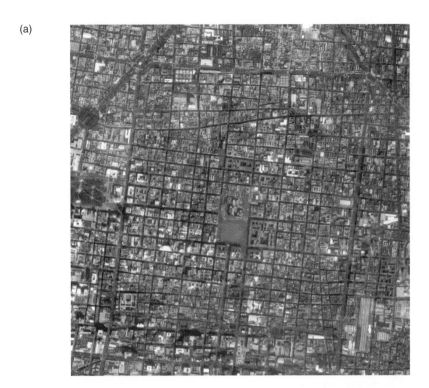

(b)

Figure 10.4(a) Air photograph of the Old City, with the central square, the Zócalo, in the centre
Source: unknown

(b) View over the old city looking eastwards towards the Zócalo. The presidential palace is on the far side of the square and the colonial age cathedral on the left
Source: author

exacerbating their harmful effects. In the Basin of Mexico, inversions occur mostly in the dry winter season in which high pressure prevails. During the wetter summer, the cloud cover and air movement generally prevent inversions.

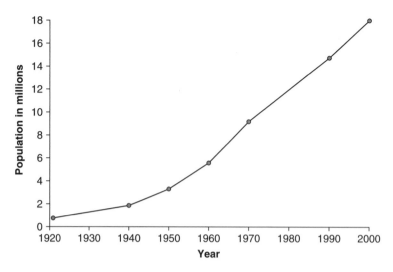

Figure 10.5   Population growth graph

Source: author

Data from Ward (1998) and INEGI (2000)

Historical considerations

Origins

In 1521, Tenochtitlán was captured, after bitter battles, by a Spanish exploratory expedition, known as the 'conquistadores', who were led by Hernán Cortés – a name now both famous and infamous throughout Mexico.

Cortés decided to build a new city on the ruins of the old, using the gridiron street pattern inherited from the Aztecs. The new city was built in what has become known as the colonial style with the grander houses built around courtyards. The Old City, or Colonial City, is now an affluent part of the modern city centre.

Recent growth

We will now skip a few hundred years and look at the more recent spectacular growth of the city. The Mexican Revolution from 1910 to about 1917 caused enormous disruption and chaos in the country, and it was not really until the 1940s that the Mexican economy, and population increase, really took off (Ward, 1998), although population growth had started long before then. The expansion in the national economy was, of course, led by Mexico City. There was huge demand for labour, and migrants began to flood in. Also, the national government encouraged this immigration. Its policies concentrated investment in Mexico City at the expense of rural areas and so contributed to both the 'push and pull' factors of migration. Despite the high migration rates, natural increase has also been a major contributor to population growth, since migrants tend to be the young, vigorous and in the most fertile age range, and after the initial surges in the city's growth, natural increase has been the main component (Ward, 1998).

In 1921, just after the revolution, the city's population was 615,000 and by 1940 was 1.64 million. Thereafter, growth was even more rapid, to 3.14 million in 1950, 5.4 in 1960 and 9.2 in 1970. After 1970, the growth rate slowed somewhat, but population still grew to 14.7 million in 1990 and 18.1 million in 2000 (INEGI, 2000; Ward, 1998). These figures provide a good guide but they are not directly comparable because, over the time period, definitions of the city have changed. This should be borne in mind if you see different population figures quoted else-where. Refer to Box 10.1 for further information on this thorny topic.

Box 10.1   Population statistics and governance problems of megacities

Determining the population of major urban areas is always fraught with difficulty, because the built-up area (or the area from which people commute into the city) rarely coincides with administrative areas, and it is administrative areas through which popu-lation statistics are compiled. Most giant cities have expanded well beyond the adminis-trative boundaries in which they began (the boundaries of the original city). Thus, there are local government organisations within the original city, and there are those in the governmental areas into which the city has spread, but there is no local government unit which has administrative powers over the whole city. Also, as the cities are continually expanding, they continually move into new administrative areas. In most cases, admin-istrative organisations with greater or lesser 'coordinating' functions (but not necessarily legal powers) across the whole city have been created. These areas usually have the word 'metropolitan' in their names, such as metropolitan area, or metropolitan zone.

If all this is a little difficult to grasp in the abstract, it will become clearer when con-sidered in relation to the map of Mexico City in Figure 10.6, and the explanation of Mexico City's unique problems outlined in Box 10.2.

Apart from difficulties with population estimation, city growth has also led to administrative problems. Box 10.2, below, describes the various administrative organisations which have an influence on the city.

Box 10.2   The government of Mexico City

Note: This box should be read in close conjunction with the map in Figure 10.6.

Mexico is a federation of states (one of which, perhaps confusingly, is the State of Mexico). Each state has considerable autonomy, similar to the USA. The national capi-tal is located in the Federal District that is not in any of the states. Subdivisions of the

(Continued)

Federal District are called Delegated Areas, roughly similar to London Boroughs. The governor of the Federal District has considerable power but, since Mexico City is the national capital, the national government also exerts a major influence.

Almost surrounding the Federal District is the State of Mexico run by a governor. Subdivisions of Mexican states are called municipalities.

As can be seen from Figure 10.6, the metropolitan area, now officially called the 'Metropolitan Zone of the Valley of Mexico', lies partly in the Federal District and partly in the State of Mexico, with one small part of it in the State of Hidalgo.

Now that the city has spilt over into the State of Mexico (see below), it is, to some extent, administered by the national government, by the Federal District government, the State Government of the State of Mexico and by the various delegated areas and municipalities. This division between administrative areas makes for problems of city definition, as described in Box 10.3, and also, as we shall see later, for difficulties with environmental management.

The population increase discussed earlier was initially mainly the result of immigration. As the city population and the density of population increased, out-migration, or decentralisation, also began. At first, it was mainly the wealthy, but later all social classes also moved outwards (Ward, 1998).

Until the 1950s the city was still contained within the Federal District, but by the late 1950s it had begun to spill over into the State of Mexico. Initially, just the nearest municipalities were affected, but decentralisation continued and is now a major trend (Pick and Butler, 1997). At the same time, new migrants were also arriving in the municipalities of the State of Mexico, adding to their growth rates and population densities. The fastest growing municipalities in the 1960s were, as might be expected, those nearest the border with the Federal District, particularly on its north-east side. These municipalities are now the most densely populated areas of the city, while in the old city centre, densities are declining as more migrants leave than arrive (Pick and Butler, 1997).

This brings us back to definitions of the city. According to Forstall et al. (2004), the 'city proper' is the older established part inside the Federal District, the 'urban area' is the continuous built-up area (which does not coincide with administrative areas), and the 'Metropolitan Zone' is the wider area that includes the whole of each delegated area or municipality that contains part of the city built-up area, with the addition of a few others because they have some functional relationship with the city (such as a high percentage of commuters. These areas are omitted from the maps in this chapter). This latter area is the one now officially known as the Metropolitan Zone of the Valley of Mexico (Box 10.2), with a population of 18.4 million at the 2000 census (INEGI, 2000). Figure 10.6 shows its administrative composition.

Different authors use the terms in various ways and often interchangeably. In this chapter, 'Mexico City' is taken to mean the whole Metropolitan Zone except where otherwise specified.

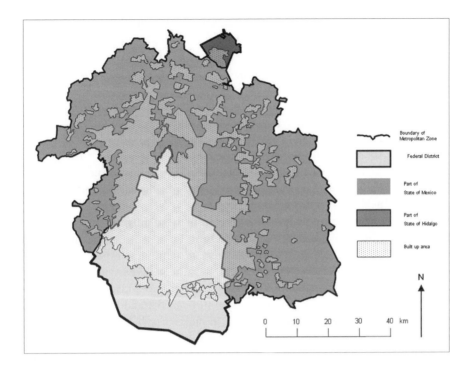

Figure 10.6 Mexico City Metropolitan Zone: administrative areas
Source: author, based on INEGI (2000)

It will be clear that the enormous, and increasing, population of the city and its physical expansion must be major factors in generating the environmental problems discussed in the following sections.

City and environment

As outlined in the introduction, one way of conceptualising the interactions between city and environment is to consider the effects of the environment on the city and vice versa. However, in the real world, the relationships are very complex and to describe (as opposed to conceptualise) them in this way would result in needless repetition. The problems will therefore be considered as a whole with the 'direction' of the influences pointed out as necessary. Also, of course, as explained in the introduction to this book, the ecological footprint of a city is much wider than the boundaries of the city, or in this case, of the Basin of Mexico. However, it is not as large as those of many western cities which, for example, export waste to the Global South.

Four main environmental problems are discussed in this section:

- tectonic instability
- air quality
- water-related problems
- solid waste.

The section concludes with a consideration of social inequalities in exposure to the effects of environmental degradation.

Tectonic instability

Attention was drawn earlier to the two great volcanoes, Popocatépetl and Iztaccíhuatl, and to the number of volcanic cones within the confines of the city. Both the big volcanoes are active but do not directly threaten the city. However, where there are volcanoes, there are earthquakes and the lake sediments under the city multiply their effects enormously. It may be that the silt liquefies in places under the pressure and compounds the damage. The result can be devastating, as in the 1985 earthquake which killed tens of thousands.

Figure 10.7 shows earthquake damage to office buildings in the city centre. There are two interesting aspects to this picture. First, it was taken in 1991, six years after the earthquake, yet the buildings are still not repaired. Secondly, though the modern office blocks are devastated, the 400-year-old colonial building in the centre is undamaged. Modern construction methods, tall buildings and government ineptitude and corruption have all been blamed for the scale of the devastation (Kandell, 1988). Although 'earthquake-proof' building technologies have been increasingly used, the amplification effects of the lake bed may well render these ineffective (Ezcurra et al., 1999). Also, earthquake proofing is difficult because the underlying geology and the propagation of seismic waves in the basin are not properly understood (Flores-Estrella et al., 2007). The effects of the earthquake are, therefore, by no means entirely due to the actions of the environment on the city.

Air quality

Mexico City's atmosphere is one of the most notorious in the world. There is very serious pollution with a complex variety of causes. There is not space here to do more than skim over some of the most important.

Anthropogenic atmospheric pollutants can be roughly divided into primary (those emitted from specific sources) and secondary (those created within the atmosphere itself from oxidation or photochemical reactions) (Molina and Molina, 2002). The most common sources of primary pollution are motor vehicles (usually the most important), and other forms of energy production and consumption, but particularly those burning fossil fuels. In addition, industries such as oil refineries and chemical plants are major contributors (Escudero, 2001; Molina and Molina, 2002).

Mexico City has these polluting activities in abundance, but perhaps the worst offender is the motor vehicle. The city is hugely congested but probably no more than any other big city. The problem here is that many of the vehicles are old and inefficient. Strict measures have been taken since 1987 to encourage newer vehicles with catalytic converters and to discourage older ones (see the section on responses to environmental problems, pp. 259–262), but despite this, vehicular pollution remains a major problem.

The high altitude of the city (see the section on location and physical characteristics, p. 238) exacerbates many of these problems because there is about 23 per cent

Figure 10.7 Earthquake damage in central Mexico City. Note the colonial era building is unharmed (though leaning a bit from earlier subsidence – see pp. 253–255)

Source: author

Figure 10.8 Popocatépetl (on the right) and Iztaccíhuatl from the city on one of those rare days

Source: martin.toluca on flickr.com

less oxygen in the air than at sea level. First, unless very carefully tuned (not often the case), vehicles burn fuel inefficiently and add to the pollution. Also, people have to inhale more air to take in sufficient oxygen and thus inhale more pollutants.

However, the best known form of air pollution in Mexico City is photochemical smog (*la contaminación*), which occurs throughout the year, though more

Figure 10.9   Smog (*la contaminación*) trapped in a temperature inversion, with Iztaccíhuatl beyond

Source of main picture: Julio Etchart (julio@julioetchart.com)

Source of insert picture: *Reforma* newspaper

commonly in winter. Very occasionally, there is a clear day and you can see the big volcanoes from the city. Not so long ago, seeing the volcanoes was normal and one of the great attractions of the city. Now if they can be seen, it excites comments in the media (Ward, 1998).

The high altitude and tropical latitude in combination facilitate the production of ozone in the presence of chemicals which, trapped in a temperature inversion, create the smog. If anyone is unclear as to what smog trapped in a temperature inversion might look like, Figure 10.9 illustrates it spectacularly. The natural conditions are not, in themselves, a hazard but become so in the presence of chemicals generated by humans.

Box 10.3   Photochemical smog

Polluted air, containing nitrogen oxides (NO_x) and volatile organic compounds (VOCs), reacts with sunlight to form photochemical smog, one of the chief components of which is ozone. Photochemical smog can cause watery eyes and difficulty in breathing, as well as damaging plants. These pollutants are emitted by a variety of processes, such as various sources of combustion and the evaporation of petrol and solvents. Photochemical smog is now recognised as a worldwide problem, particularly in places with high NO_x and VOC emissions, which are trapped by temperature inversions and exposed to sunlight. (Molina and Molina, 2002)

Finally, it is necessary to be aware that 'air pollution' is a socio-political construct. Of course, the pollution is there in the atmosphere, but it is a political decision as to when the level of contaminants is defined as pollution (Escudero, 2001), and therefore what measures may be taken to combat it. These issues are discussed later when considering environmental policymaking.

Water (incoming and outgoing)

From everything considered so far about Mexico City, it will be clear that:

- there is a huge demand for water
- also, therefore, a huge quantity of waste water is disposed of
- serious environmental consequences are likely.

Mexico City consumes more than 300 litres of water per capita per day, which is more than many European cities (Ezcurra et al., 1999 and see Chapter 6, p. 122). However, the report (National Academy of Sciences, 1995) that is described below as authoritative (having been compiled by American and Mexican water experts) is questioned by Castro (2004) for favourably comparing Mexico City's consumption with the average of the United States, which is around 650 pclpd. Also, of course, averages are deceptive. While certain social groups consume well above the average, others, hugely greater in number, have serious deficiencies in water availability (as discussed under *Inequality* on p. 253).

In addition, the nature of the physical environment (see p. 237) creates problems both for supply and disposal which are greatly exacerbated by anthropogenic factors. Further, the city's water distribution system suffers from serious leakage problems. Ezcurra et al. (1999) suggest 25 per cent is lost but Castro (2004), citing the director of the Federal District Water Commission, puts it at between 30 and 50 per cent.

Water supply

Mexico City's water comes from two main sources: the aquifer (Box 10.4) under the basin, and from ground and surface water drawn from the Lerma and Cutzamala basins (Figure 10.11), outside the Basin of Mexico.

From about the middle of the 19th century, water has been drawn from the aquifer underlying the city but, as the population increased, so did the need for more water and the extraction of groundwater has increased enormously. Very few wells now have artesian pressure and most of the water has to be pumped. At present, about 70 per cent of the city's water comes from within the basin and 30 per cent from outside (Gonzalez-Moran et al., 1999).

Box 10.4   Artesian wells

An 'aquifer' is a layer of porous rock which permits water to pass through it. If the aquifer has non-porous rock above and below it, and is folded into the shape of a basin, the water is trapped in the lower part of the basin. Rain or melting snow enters the aquifer where it outcrops at the surface.

(Continued)

'Artesian wells' are drilled from the surface into the aquifer. If the level of water in the aquifer is higher than the wells, 'artesian pressure' will force the water out. However, if the amount of water extracted is greater than its replenishment by rain water, the artesian pressure gradually decreases and the water has to be pumped out.

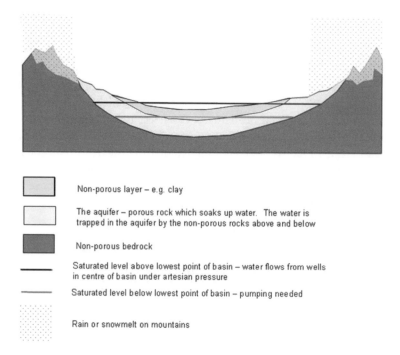

Non-porous layer – e.g. clay

The aquifer – porous rock which soaks up water. The water is trapped in the aquifer by the non-porous rocks above and below

Non-porous bedrock

Saturated level above lowest point of basin – water flows from wells in centre of basin under artesian pressure

Saturated level below lowest point of basin – pumping needed

Rain or snowmelt on mountains

Figure 10.10   Schematic diagram of an artesian basin
Source: author

However, as demand for water and depletion of the aquifer continue to increase, there will inevitably be supply problems. It is clear from decreasing artesian pressure that the aquifer is being overexploited, but no one knows by how much (Ezcurra et al., 1999; Gonzalez-Moran et al., 1999). The problem is that the characteristics of the aquifer are not properly understood. Indeed, one of the recommendations from an authoritative report (National Academy of Sciences, 1995) states: 'A long-term research program for determining the hydrologic, physical, chemical, and biological characteristics of the aquifers in the Basin of Mexico should be developed' (p. 79).

The problem is not confined to the aquifer. The Lerma and Cutzamala basins, referred to above, are also suffering from scarcity and depletion of water which is needed in those areas.

Both sources of water have to be pumped – from the aquifer because of loss of artesian pressure and, particularly, from outside the basin to raise the water

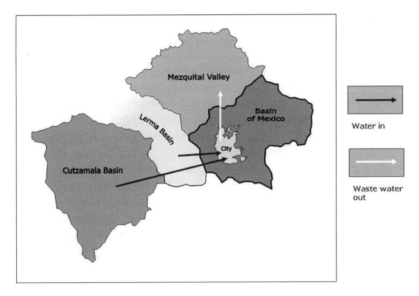

Figure 10.11 The Lerma and Cutzamala basins and the Mezquital Valley: locations
Source: author

sufficiently to cross the watershed enclosing the basin. A huge amount of energy is needed to raise the water to the requisite levels, which is both inefficient and expensive (Ezcurra et al., 1999).

Waste water

Waste water, and excess surface water during the rainy season (the city suffers severely from surface floods during storms), is exported from the basin by a combination of surface channels and deep drainage tunnels, to which approximately 75 per cent of the population is connected. However, in addition to domestic waste the tunnel drainage system is also used to dispose of untreated industrial liquid waste. The tunnels are supposed to be watertight, but there is obviously a risk of leakage in such an unstable geological area. Although there are water treatment plants, they manage to treat only about 7 per cent of total waste water (Ezcurra et al., 1999). The exported waste passes northwards out of the basin and eventually into the Mezquital Valley (Figure 10.11) in the state of Hidalgo.

The surface canals originally sloped downwards from the central city to tunnels through the mountains. However, subsidence (see pp. 251–253) has meant that the wastewater level, just as with incoming water, now has to be raised by pumping to a level sufficient to permit its export across the watershed.

Water-related environmental issues

The nature of Mexico City's water supply and disposal leads to major environmental difficulties. There are four main areas:

- supply and disposal costs (both economic and social)
- contamination

- subsidence
- inequality.

Supply and disposal costs   Tortajada and Castelán (2003) point out that the costs of constructing the necessary infrastructure for supply and disposal have run to hundreds of millions of US dollars. Also, the government has concentrated on such engineering projects, rather than developing appropriate environmental, economic and social policies, with consequent severe costs in health and environment (refer to the following sections).

The cost in energy supply of pumping water within, into and out of the basin, as described above, is extremely high, and with the increasing growth of the city, more and more pumping will be required until, perhaps, it will become no longer economically viable. Also, of course, there are the environmental costs of generating the necessary electricity, though Mexico does have substantial hydro-electric sources and two nuclear plants. Legorreta et al. (1997) claim that the energy needed to raise the water from the Cutzamala Basin is equivalent to that consumed by the city of Puebla, with a population of over 8 million. As Ezcurra et al. (1999, p. 147) say, 'The energy costs involved in pumping water into, within and out of the Basin of Mexico . . . appear at present to be one of the most important factors limiting the growth of Mexico City'.

Further, pumping imposes severe social costs in those areas outside the basin from which water is drawn and to which it is exported. In the Lerma and Cutzamala basins, there is already considerably reduced water available for irrigation and electricity generation, and it has also been suggested (Rudolph et al., 2006) that excessive extraction may result in a reduction in groundwater quality. In Hidalgo, the untreated waste water is used for irrigation and thus contaminates agricultural land and produce (see next section).

Montezuma's revenge   Visitors to Mexico City often refer to 'Montezuma's revenge', the stomach ailment that assails many newcomers to the city, caused by the drinking water or unwashed salad or fruit. Montezuma, or Moctezuma in Spanish, was the Aztec emperor ousted by the conquistadors. The facetious name suggests nothing very much more than a short bout of diarrhoea and sickness. However, it can be much more debilitating and long term, as this author can testify. Indeed, Ezcurra et al. (1999) point out that diseases associated with drinking water are among the most common causes of death in Mexico. Both chemical and bacterial contaminants are found in groundwater, but it is not entirely clear how they enter the aquifer, though there are a number of studies investigating the processes (Ezcurra et al., 1999; Mazari-Hiriart et al., 2005). One possible source is the deep water drainage tunnels. These have been overused to the extent that they have not been inspected for 10 years. It is possible that given considerable earth movements (particularly subsidence, see the section on the sinking city below), leakage has occurred and inspection is urgently required (Cisneros-Iturbe and Dominguez-Mora, 2005).

It is not, however, just the drinking water. Much of the waste water, as has been seen above, is an untreated mixture of domestic and industrial waste that is used for irrigation in the Mezquital Valley in the state of Hidalgo. This, of course, leads

The circled areas are not easy to see but show steps down from street level which were not part of the original design.

Cars parked next to the building are on a slope caused by the sinking

Figure 10.12 Palacio de Bellas Artes
Source: © George and Audrey DeLange

to the crops themselves becoming contaminated with obvious consequences for the consumers, as well as being a health hazard for the farmers in the valley (Tortajada, 2006). Although irrigating high-value crops for human consumption with waste water is now banned in the area, desperate farmers feel they have to resort to illegal methods to make a living (Ezcurra et al., 1999). Also, surface water used for irrigation in the south of the city has been shown to contain bacterial contamination above the recommended level for irrigation water (Solis et al., in press). It is clear, therefore, why it is necessary to wash the salad. Less clear is what you wash it *with*.

The sinking city   One of the most spectacular consequences of building on soft clay and of excessive water extraction is subsidence, which is made even more bizarre by different rates of subsidence of different objects, or in different parts of the city. Figures 10.12, 10.13 and 10.14 illustrate some of these effects.

One of Mexico City's best known landmarks, the Palacio de Bellas Artes, is sinking. The weight of the building has compressed the soft clays of the former lake bed. Figure 10.12 shows some of the ways in which the sinking can be identified.

Figure 10.13 shows the opposite effect – the road has sunk relative to the building. This effect is created where the building has been constructed on deep piles and has not subsided while the land around it has.

The pictures illustrate relative subsidence of a metre or so and unstable buildings (Figure 10.14) caused mainly by sinking in the soft sand of the former lake bed. What none of the pictures can show, however, is that the whole city has subsided by around 7.5 metres in the central area in the last 100 years, with some

Figure 10.13   Unknown building, 1977

Source: Geotimes (www.geotimes.org). Picture supplied courtesy of Thomas Holzes of the USGS

Figure 10.14   Leaning churches in Mexico City

Source: www.unep.or.jp/.../IMG/photo45Mexicocity.gif

Division of Technology, Industry and Economics, United Nations Environment Programme

areas to the south of the city reaching 15 metres (Gonzalez-Moran et al., 1999). This overall sinking is the result of water extraction from the aquifers, and is responsible for effects which are less visible and spectacular, but perhaps more important, than surface instability.

The surface wastewater drains (p. 249) once sloped downwards from the central city to the tunnels through the mountains but no longer do so, and pumping waste water up to the levels of the main channels has become necessary. The largest channel, the Gran Canal, has completely lost inclination for its first 20 km,

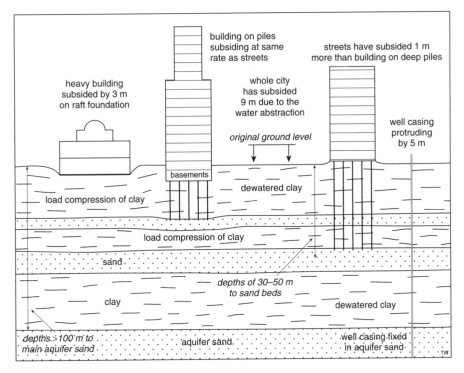

Figure 10.15 Mexico City: subsidence typology
Source: Waltham, 2002

and a new gravity-fed tunnel has now been constructed, 20 m below the surface, to take stormwater run-off to it at a point where it is still sloping (Tortajada, 2006). The deep drainage tunnels are also possibly affected (see p. 249), even though they were constructed at a level which was supposed to be deep enough not to be affected by subsidence (Tortajada, 2006). Also, Mexico City has a very efficient subway system, but its operating costs are greatly increased because the lines have to be realigned every six months (Gonzalez-Moran et al., 1999).

Water extraction in the central area was discontinued in 1960 and subsidence was greatly decreased though not stopped entirely. However, it has increased greatly in some outer areas to the east and south, and subsidence there has increased concomitantly with severe environmental effects (Gonzalez-Moran et al., 1999; Lesser and Cortes, 1998).

Figure 10.15 shows a typology of differential rates of sinking in the city. Note the overall subsidence, within which the differential sinking takes place.

Inequality   The illustration in Figure 10.16 is all too familiar in the poorer parts of Mexico City. It shows a resident collecting water from an official distributor at a distribution point.

In general, the poorer parts of the city lie to the east as shown in Figure 10.17. These areas are the driest on the surface, yet are those where most water is currently being extracted from the aquifers, and they are also the most susceptible

Figure 10.16   Water delivery by an official Mexico City water distributor, March 2006
Source: Planetsave, PO Box 11282, Portland, ME, USA, 04104

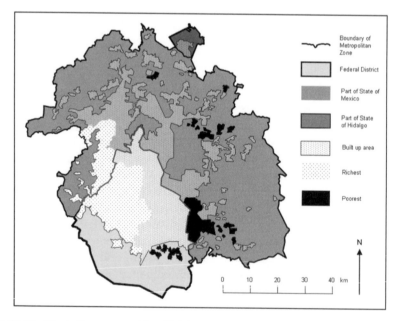

Figure 10.17   The built-up areas of the 15 richest and 15 poorest delegated areas and
municipalities in the Metropolitan Zone

Note: The richest are defined as the 15 delegated areas and municipalities with the highest percentages of the employed population earning more than 10 times the minimum salary. The poorest
are defined as the 15 delegated areas and municipalities with the highest percentages of the
employed population earning less than twice the minimum salary (used by Ezcurra et al. (1999) to
identify 'the poverty level').

Source: author, based on census data (INEGI, 2000)

Table 10.1   Domestic water supply in the Metropolitan Zone

	Number of Households	Percentage of Households by Access to Piped Water			
		Inside the house	Inside the property but outside the house	From taps or hydrants in the vicinity	No access
Metropolitan Zone	4,846,896	64.11	28.81	2.46	3.76
Fifteen richest delegated areas and municipalities	1,708,807	76.78	19.56	1.32	1.51
Fifteen poorest delegated areas and municipalities	1,119,388	56.45	37.55	1.98	3.26
Individual municipality with highest in-house supply	113,741	96.53	2.23	0.22	0.28
Individual municipality with lowest in-house supply	99,372	20.00	64.30	3.43	11.45

Note: The 15 richest and poorest delegated areas and municipalities are those identified in Figure 10.17.
Source: author, based on census data (INEGI, 2000)

to contamination. Thus, in terms of water, the poor suffer disproportionately from unavailability, subsidence and disease.

Table 10.1 demonstrates some of the variability in water supply. Although the overall coverage of piped water is quite high, the difference between the richest and the poorest is plain. Consider, for example, the high percentage of households which do not have piped water in the house. Even in the richest 15, it is only three-quarters, while in the poorest, it is barely over half.

The notion of piped water in the property but outside the house may be unfamiliar. The hosepipe in the foreground of Figure 10.18 gives a clue as to what it implies.

However, not so obvious is the plight of the 3.76 per cent of households in the Metropolitan Zone which do not have access to piped water at all. The resident in Figure 10.16 is among the more fortunate of the poorest, in that he is receiving an official supply, but many are forced to rely on buying water from private sellers, usually at highly inflated prices. Tortajada (2006) suggests that water in 200-litre containers from private trucks costs 500 per cent more than piped water which represents between 6 and 25 per cent of people's income. Castro (2004) claims that the number of people so affected runs into millions. Perhaps 3.76 per cent may not seem a lot but it still amounts to nearly 200,000 out of 4.85 million households. With a total population of over 18 million in those households, Castro's figure does not seem unreasonable.

Although demand for water is very high, there is not an overall shortage. On p. 247, it was observed that the average consumption in the city was over

Figure 10.18   The 'courtyard' of a home in Nezahualcóyotl, a poor area in the east of the city
Source: author

300 litres per capita per day (pclpd). However, wealthier households may consume over 600 pclpd, while in the poorer areas, it is perhaps 20 litres, and in the
most deprived localities within the poorest delegated areas and municipalities,
pclpd may be as low as four litres (Castro, 2004; Tortajada, 2006).

Castro's article also draws attention to the numerous public protests, demonstrations and even violence (not always perpetrated by the demonstrators) which
have taken place in relation to the gross discrepancies in water supply just
described. Not surprisingly, the great majority of such protests took place in the
poorer parts of the city. Mexico has a long history of public demonstrations, many
of them successful but others suppressed. The contest between those demanding
water as a public right and those supporting privatisation continues. Figure 10.19
shows a peaceful demonstration in March 2006 during the World Water Forum
which was held in Mexico City. The banner reads:

> **WATER** is a **right**
> not a commodity
> **NO** to privatisation

Figure 10.19   Demonstration during the World Water Forum, Mexico City, March 2006
Source: www.uusc.org/blog/hotwire/2006_03_01_archive.html

Subsidence is another problem that disproportionately affects poorer areas. The differences are not as great as for availability because, as has been seen above (pp. 251–253), the central, more affluent, parts of the city have severe problems. However, as Gonzales-Moran et al. (1999) have observed, parts of the east and south, such as Chalco, have experienced subsidence of up to 15 metres and, they point out, residential construction methods in poor areas are likely to be substandard and will thus be more susceptible to damage.

Poor water quality, with its attendant health problems, is also found all over the city but, like the other negative effects already discussed, contamination is concentrated mostly in the east (Ezcurra et al., 1999; Gonzalez-Moran et al., 1999). There is justifiable widespread suspicion of the water throughout the city, and because of this, large numbers of people drink only bottled water. As a result, Mexico City has become the world's second largest consumer of bottled water (Rodwan, 2004). Although the problem is city-wide, it is clearly the poor who will be hardest hit. Not only do many of them have to pay heavily for ordinary water, but to stay healthy, they really ought to buy bottled water as well. Because of the inability to afford the additional expenditure, added to the higher levels of contamination in those same areas, it is not surprising that the poorer localities are the worst affected by water-related illnesses.

Solid waste

At this point, it will be instructive to stop reading this chapter for a few minutes and refer back to the introductory paragraphs of Chapter 7, on waste (pp. 150–151). How easy it is for one person to generate waste without really noticing! Now multiply one person's discards by 18 million and you begin to see Mexico City's problem. And that is only personal rubbish. Consider all the other types of waste outlined in Table 7.1, p. 152. Then add in a comparative lack of awareness in Mexico and it is clear that there is a major problem.

The city generates approximately 26,000 tonnes of municipal solid waste (MSW; see also Chapter 7) per day, of which 76 per cent ends up (after several stages described below) in two final disposal sites or other managed dumps, while the other 24 per cent goes to open dumps scattered throughout the metropolitan area, but concentrated particularly in the east (Castillo Berthier, 2003; de la Rosa et al., 2006). In the Mexican classification, MSW includes domestic, commercial, office and institutional waste but not industrial waste (Ezcurra et al., 1999).

In addition to the sheer volume of rubbish, its composition adds to the problem. Some 47 per cent of MSW is organic, and therefore biodegradable, matter (compared with about 30 per cent in the USA) which could be used as compost (but is not, to date). As it degrades, it produces gases. One of these is methane, which is a greenhouse gas and the others are VOCs (see page 248 in this chapter) which contribute to photochemical smog. The idea of collecting these gases as a source of energy has been promulgated for a long time as a viable possibility (de la Rosa et al., 2006; Mulás, 1995). For now, however, they remain another environmental hazard.

Organic waste is not recyclable (except as in the possibilities mentioned above). This means that only about half of the city's waste is potentially recyclable. Although this is a small proportion, everything else (apparently paradoxically) that can be recycled *is* recycled – though most of it not through official channels. There are three recently established recycling plants in the Federal District, though none in the State of Mexico. However, although the Federal District operates the plants, the organisation of the work and the sale of recycled products is, effectively, outside their control and in the hands of powerful bosses who control a workforce with absolute authority. The bosses decide who is paid and how much, and organise the sale of recycled materials. And this is in *official* recycling plants! These plants recover only about 10 to 13 per cent of the waste they receive and the remainder goes on to landfill. But that is not the end of recycling.

Look at Figure 10.20. This shows three *pepenadores*, garbage pickers, in the background on a dump on the outskirts of Mexico City. In the foreground, there is a structure. This is one of many where the *pepenadores* live.

These are the poorest of the poor but scavenging provides a living. Moreover, it is a highly organised living. Just as the workforce in the recycling plant is organised by bosses, so are the *pepenadores*. The bosses are ruthless and exercise total control but they do guarantee enough money to live on – just. Everything that can possibly be sold is collected and delivered, via the bosses who control the price, in an upward chain to the final users. Across the metropolitan area there are still thousands of people living and scavenging on dumps. According to Castillo Berthier (2003), until the early 1980s, there were 10,000 people living on a single notorious dump.

The scavenging process is effective but it is not very efficient. Mexico City's environment would be better served if recyclable materials were separated at source, processed in more recycling plants and distributed to users directly. There are moves in this direction, but they are bitterly resisted by the *pepenadores*

Figure 10.20   Garbage pickers and their shacks on a dump on the outskirts of the city
Source: author

because they would deprive them of their living. It may be thought that these workers would have little influence but, because of the way grassroots politics works in Mexico, they do have a voice which can be quite powerful.

So far, we have considered only municipal waste. However, the greatest threat from waste to Mexico's environment comes from industrial waste, particularly hazardous industrial waste (see Chapter 7). There are very few public sites for the disposal of hazardous waste in the whole of Mexico and none in Mexico City (Ezcurra et al., 1999). Given the concentration of Mexico's industrial capacity in the city, this is extraordinary and environmentally very dangerous. As has been seen earlier, much liquid industrial waste is discharged into public sewers and the situation for solids is not much better. Certainly, a considerable quantity arrives in the main landfills, which were not designed for the purpose and present a major risk of contamination, either atmospheric or through leaching into the groundwater which, of course, is a major source of water for the city. A study (Gonzalez-Moran, 2002) at one such major landfill suggests that contamination is indeed occurring at wells in the vicinity. Even more dangerous is hazardous waste that is deposited in the open, uncontrolled, dumps where there is not even minimal protection from leaching.

Responses to environmental problems

It is important not to give an impression that is too negative for Mexico City as a whole. In many ways, the city is vibrant and successful, but the environment, with which this book is concerned, is clearly in a parlous state, teetering on the edge of unsustainability, as suggested by the quotation at the beginning of the chapter. This does not mean, however, that nothing is being done about it. Over the last 15 years, the metropolitan area has made great strides in combating environmental problems. Regulations are in place to control emissions from all sources (Molina and Molina, 2002), water supply and disposal, and solid waste. There is a plethora of organisations, large and small, including governmental, non-governmental, voluntary, single issue and many others. Some of these are discussed in this section, though it is not possible in the space available to give more than a general overview, with a focus on some of the more important.

Until about the early 1980s, the environment was not really being taken all that seriously, although awareness was gradually increasing and there had been some legislation. Then, in 1982, the central government realised that action was needed and a quite powerful new environment ministry was created. Initially, not much action was taken, mainly because of a national economic crisis and the devastating earthquake of 1985 (see p. 244). However, air pollution continued to become even more alarming and a series of anti-pollution measures was introduced in 1986, mostly related to the big political issue – air quality – but including the creation of managed landfills, which were lined with clay to prevent leaching and periodically covered with clay as well.

Roughly parallel with central government initiatives, environmental action (again mostly air pollution-related) was being taken by the governments of the Federal District and the State of Mexico. However, these were badly coordinated and it was really left to central government to deal with metropolitan issues. Indeed, as indicated earlier on p. 242, coordination between the three main governments (Federal, Federal District and the State of Mexico) is poor, and metropolitan-wide coordinating institutions, though they exist, remain very weak (Ortega-Alcazar, 2006). For example, the Metropolitan Environmental Commission is intended to be an important organisation to coordinate the policies of the various governments, but it does not have its own budget nor a properly identified structure (Molina and Molina, 2002). Also, at the time of writing, the three governments were controlled by three competing political parties which does not help the cause of cooperation. The absence of powerful metropolitan-wide bodies is clearly a major drawback in dealing with, for example, atmospheric pollution, which affects the whole of the basin, or the transport network, which itself is integrated but is shared between the State of Mexico and the Federal District. Transport and its attendant congestion is, as seen earlier, the main contributor to poor air quality.

Partly in response to this ineffectiveness, in 1987, one small environmental group had an idea that led to Mexico's best known (and perhaps most notorious) anti-pollution measure, known as *Hoy no circula*, or 'don't drive today'. They persuaded drivers throughout the metropolitan area not to use their cars for one day a week. In 1989, the idea was taken up by government and became mandatory. To begin with, it had wide support and pollution declined accordingly. Unfortunately, the restriction more or less coincided with an economic boom, and people responded by buying two cars, so they could always drive one and frequently both on the other days of the week. Between 1989 and 1994, average car sales doubled, resulting in a huge increase in the numbers of cars on the road, and petrol consumption rose by up to 30 per cent (Ezcurra et al., 1999). The

Figure 10.21   Stickers indicating the day not allowed to drive during the heyday of *hoy no circula* in 1992
Source: author

government responded by introducing measures to reduce harmful emissions, and cars complying with the new regulations were allowed to drive every day but, at the same time, older, less efficient cars became subject to no driving for two days in 1995. *Hoy no circula* is still in operation but in a much modified form, changing gradually as new technological restrictions have been, and continue to be, imposed.

We have seen that environmental issues related to water are central to Mexico City's problems and, indeed, according to Ezcurra et al. (1999), water supply and waterborne disease may well be the most important of all. At the national level, there is a determination to develop a sustainable approach to the planning and management of water resources. However, coordination between the two main city authorities and local organisations is, once again, very complex . One important example of an attempt to involve all interested parties is the creation of River Basin Councils, including one for the Valley of Mexico (Paredes, 1997). These are intended to develop policies resulting from full participation by all stakeholders, from local people to the CNA, the National Water Commission. The intention is very laudable, but the wide variation in size, funding and motivation of the many partners, together with the opposition of vested interests, means that progress is very slow and cumbersome.

Finally, Mexico as a nation is committed to advance sustainability and environmental awareness in conjunction with international bodies such as UNEP (the United Nations Environment Programme). Perhaps most important, however, is the North American Agreement for Environmental Cooperation which accompanies the North American Free Trade Agreement (NAFTA), through which Mexico is obliged to bring its environmental quality broadly into line with the United States and Canada. There are some, however, who suggest that membership of NAFTA *increases* the risk of environmental degradation. These issues are concisely but fully examined in a short article by Kevin Gallagher (2004).

Summary

At this point, it would be a good idea to look back at the learning outcomes listed at the start of the chapter, and check whether they have been attained, then, perhaps, read parts again to consolidate the learning process.

It is clear that, despite the measures outlined in the previous section, the environmental situation in the basin remains critical. The physical setting changes very little, but population continues to grow rapidly with its attendant environmental consequences, such as the increase in motor vehicles (and their emissions), demand for water, and waste to get rid of. The economy of the city continues to burgeon, which is of course in many ways a major asset. Moreover, most of the growth is in the less polluting service sector, but the city still has the principal concentration of industry in the country, which, despite controls, contributes massively to environmental degradation.

Of the many problems considered, the best known is atmospheric pollution, and many measures are in place to address it. The most serious, however, from the perspective of sustainability and continued city growth, is water supply. At the time of writing, the authorities cited in the text are agreed that the environment in Mexico City and the Basin of Mexico has reached a tipping point. It now remains to be seen whether water supply, and all the other problems, can be overcome in time to prevent the city going over the edge.

References

Castillo Berthier, H. (2003) 'Garbage, work and society', *Resources, Conservation and Recycling*, 39(3): 193–210.

Castro, J.E. (2004) 'Urban water and the politics of citizenship: the case of the Mexico City Metropolitan Area during the 1980s and 1990s, *Environment and Planning A*, 36(2): 327–46.

Cisneros-Iturbe, H.L. and Dominguez-Mora, R. (2005) 'Strategy to allow the inspection of the deep drainage system of Mexico City', *IAHS-AISH Publication*, 293: 212–20.

de la Rosa, D.A., Velasco, A., Rosas, A. and Volke-Sepulvéda, T. (2006) 'Total gaseous mercury and volatile organic compounds measurements at five solid waste disposal sites surrounding the Mexico City Metropolitan Area', *Atmospheric Environment*, 40: 2079–88.

Escudero, C.N. (2001) *Urban Environmental Governance: Comparing Air Quality Management in London and Mexico City*. Aldershot: Ashgate.

Ezcurra, E., Mazari-Hiriart, M., Pisanty, I. and Aguilar, A.G. (1999) *The Basin of Mexico: Critical Environmental Issues and Sustainability*. Tokyo: United Nations University Press.

Flores-Estrella, H., Yussim, S. and Lomnitz, C. (2007) 'Seismic response of the Mexico City Basin: a review of twenty years of research', *Natural Hazards*, 40(2): 357–72.

Forstall, R.L., Greene, R.P. and Pick, J.B. (2004) 'Which are the largest? Why published populations for major world urban areas vary so greatly', in *City Futures: an international conference on globalism and urban change*, University of Illinois at Chicago, July.

Gallagher, K.P. (2004) *Free Trade and the Environment: Mexico, NAFTA, and Beyond*. Palo Alto, CA: Stanford University Press.

Gonzalez-Moran, T. (2002) 'Functional model of groundwater alteration in the vicinity of a landfill, Santa Catarina, Chalco, Mexico', *Revista Geofisica*, 57: 111–24.

Gonzalez-Moran, T., Rodriguez, R. and Cortes, S.A. (1999) 'The Basin of Mexico and its metropolitan area: water abstraction and related environmental problems', *Journal of South American Earth Sciences*, 12(6): 607–13.

Hall, T. (2006) *Urban Geography*. Abingdon: Routledge.

INEGI (2000) 'XII censo general de población y vivienda 2000', http://www.inegi.gob.mx

Kandell, J. (1988) *La Captital*. New York: Henry Holt.

Kasperson, J.X., Kasperson, R.E. and Turner II, B.L. (1999) 'Preface', in E. Ezcurra, M. Mazari-Hiriart, I. Pisanty and A.G. Aguilar (eds), *The Basin of Mexico: Critical Environmental Issues and Sustainability*. Tokyo: United Nations University.

Legorreta, J., Contreras, M.C., Flores, M.A. and Jimenez, N. (1997) 'Agua y mas agua para la ciudad', *Ecologica* (*La Jornada* supplement), July.

Lesser, J.M. and Cortes, M.A. (1998) 'Land settling in Mexico City and its implications in the drainage system', *Ingenieria Hidraulica en Mexico*, 13(3): 13–18.

Mazari-Hiriart, M., Lopez-Vidal, Y., Ponce-de-Leon, S., Calva, J.J., Rojo-Callejas, F. and Castillo-Rojas, G. (2005) 'Longitudinal study of microbial diversity and seasonality in the Mexico City metropolitan area water supply system', *Applied and Environmental Microbiology*, 71(9): 5129–37.

Molina, L.T. and Molina, M.J. (eds) (2002) *Air Quality in the Mexico Megacity: An Integrated Assessment*. Dordrecht: Kluwer Academic.

Mulás, P. (1995) 'Energía y desarrollo sostenible: el caso de México', in L. García-Colín and M. Bauer (eds), *Energía, ambiente y desarrollo sustentable: El caso de México*. Mexico D.F.: El Colegio Nacional and UNAM.

National Academy of Sciences (1995) *Mexico City's Water Supply: Improving the Outlook for Sustainability*. Washington, D.C.: National Academy Press.

Ortega-Alcazar, I. (2006) 'Mexico City: growth at the limit?', http://www.urban-age.net/03_conferences/conf_mexicoCity.html

Paredes, A.J. (1997) 'Water management in Mexico: a framework', *Water International*, 22(3): 135–9.

Pick, J.B. and Butler, E.W. (1997) *Mexico Megacity*. Oxford: Westview Press.

Rodwan, J.G. (2004) 'Bottled water 2004: US and international statistics and developments', *Bottled Water Reporter*, April/May.

Rudolph, D.L., Sultan, R., Garfias, J. and McLaren, R.G. (2006) 'Significance of enhanced infiltration due to groundwater extraction on the disappearance of a headwater lagoon system: Toluca Basin, Mexico', *Hydrogeology Journal*, 14(1–2): 115–30.

Solis, C., Sandoval, J., Perez-Vega, H. and Mazari-Hiriart, M. (in press) 'Irrigation water quality in southern Mexico City based on bacterial and heavy metal analyses', *Nuclear Instruments and Methods in Physics Research Section B: Beam Interactions with Materials and Atoms*, in press, corrected proof.

Tortajada, C. (2006) 'Water management in Mexico City Metropolitan Area', *International Journal of Water Resources Development*, 22(2): 353–76.

Tortajada, C. and Castelán, E. (2003) 'Water management for a megacity: Mexico City Metropolitan Area', *Ambio*, 32(2): 124–9.

Waltham, T. (2002) 'Sinking cities', *Geology Today*, 18(3): 95–100.

Ward, P.M. (1998) *Mexico City: Revised Second Edition*. Chichester: Wiley.

Index